Design Justice

Information Policy Series

Edited by Sandra Braman

The Information Policy Series publishes research on and analysis of significant problems in the field of information policy, including decisions and practices that enable or constrain information, communication, and culture irrespective of the legal siloes in which they have traditionally been located, as well as state-law-society interactions. Defining *information policy* as all laws, regulations, and decision-making principles that affect any form of information creation, processing, flows, and use, the series includes attention to the formal decisions, decision-making processes, and entities of government; the formal and informal decisions, decision-making processes, and entities of private and public sector agents capable of constitutive effects on the nature of society; and the cultural habits and predispositions of governmentality that support and sustain government and governance. The parametric functions of information policy at the boundaries of social, informational, and technological systems are of global importance because they provide the context for all communications, interactions, and social processes.

Virtual Economies: Design and Analysis, Vili Lehdonvirta and Edward Castronova

Traversing Digital Babel: Information, e-Government, and Exchange, Alon Peled

Chasing the Tape: Information Law and Policy in Capital Markets, Onnig H. Dombalagian

Regulating the Cloud: Policy for Computing Infrastructure, edited by Christopher S. Yoo and Jean-François Blanchette

Privacy on the Ground: Driving Corporate Behavior in the United States and Europe, Kenneth A. Bamberger and Deirdre K. Mulligan

How Not to Network a Nation: The Uneasy History of the Soviet Internet, Benjamin Peters

Hate Spin: The Manufacture of Religious Offense and Its Threat to Democracy, Cherian George

Big Data Is Not a Monolith, edited by Cassidy R. Sugimoto, Hamid R. Ekbia, and Michael Mattioli

Decoding the Social World: Data Science and the Unintended Consequences of Communication, Sandra González-Bailón

Open Space: The Global Effort for Open Access to Environmental Satellite Data, Mariel John Borowitz

You'll See This Message When It Is Too Late: The Legal and Economic Aftermath of Cybersecurity Breaches, Josephine Wolff

The Blockchain and the New Architecture of Trust, Kevin Werbach

Digital Lifeline? ICTs for Refugees and Displaced Persons, edited by Carleen F. Maitland

Designing an Internet, David D. Clark

Reluctant Power: Networks, Corporations, and the Struggle for Global Governance in the Early 20th Century, Rita Zajácz

Human Rights in the Age of Platforms, edited by Rikke Frank Jørgensen

The Paradoxes of Network Neutralities, Russell Newman

Zoning China: Online Video, Popular Culture, and the State, Luzhou Li

Design Justice: Community-Led Practices to Build the Worlds We Need, Sasha Costanza-Chock

Design Justice

Community-Led Practices to Build the Worlds We Need

Sasha Costanza-Chock

The MIT Press
Cambridge, Massachusetts
London, England

The open access edition of this book was made possible by generous funding from Knowledge Unlatched.

Knowledge Unlatched

This book was set in ITC Stone Serif Std and ITC Stone Sans Std by Toppan Best-set Premedia Limited. Printed and bound in the United States of America.

Library of Congress Cataloging-in-Publication Data

Names: Costanza-Chock, Sasha, 1976- author.
Title: Design justice : community-led practices to build the worlds we need / Sasha Costanza-Chock.
Description: Cambridge, MA : The MIT Press, 2020. | Series: Information policy | Includes bibliographical references and index.
Identifiers: LCCN 2019015279 | ISBN 9780262043458 (hardcover : alk. paper)
Subjects: LCSH: Design--Social aspects. | Social justice.
Classification: LCC NK1520 .C675 2020 | DDC 745.4--dc23 LC record available at https://lccn.loc.gov/2019015279

10 9 8 7 6 5 4 3 2

For Allied Media Projects, a collaborative laboratory of media-based organizing whose transformative impacts will ripple outward for generations.

For Mother Cyborg and the Detroit Community Technology Project, visionaries of digital stewardship, DiscoTechs, and building the worlds we need.

For two-spirit people, who survived centuries of settler colonialism and are still here, reclaiming their rightful places beside the council fires.

Contents

Acknowledgments

This book reflects the labor of a great many people. First, there would be no design justice theory or practice as we know it without years of work from the many Design Justice Network organizers, especially Una Lee, Victoria Barnett, Wes Taylor, Carlos (L05) Garcia, Nontsikelelo Mutiti, Adrienne Gaither, Taylor Stewart, Ebony Dumas, Danielle Aubert, Victor Moore, and Gracen Brilmyer, as well as the hundreds of people who have organized and taken part in design justice workshops at the Allied Media Conference since 2016. The authors of the first version of the Design Justice Network Principles are Una Lee, Jenny Lee, Melissa Moore, Wesley Taylor, Shauen Pearce, Ginger Brooks Takahashi, Ebony Dumas, Heather Posten, Kristyn Sonnenberg, Sam Holleran, Ryan Hayes, Dan Herrle, Dawn Walker, Tina Hanaé Miller, Nikki Roach, Aylwin Lo, Noelle Barber, Kiwi Illafonte, Devon De Lená, Ash Arder, Brooke Toczylowski, Kristina Miller, Nancy Meza, Becca Budde, Marina Csomor, Paige Reitz, Leslie Stem, Walter Wilson, Gina Reichert, and Danny Spitzberg. Nor would this book be possible without everyone from the Tech for Social Justice Project (T4SJ), the #QTPower crew, and my Allied Media fam. Diana Nucera, especially, you're a guiding light in the community technology movement: House of Cyborg forever! Thanks also to all the staff, research assistants, students, and community partners from the Collaborative Design Studio, too numerous to list here (explore https://codesign.mit.edu for more info).

Next, profound thanks to my editors at the MIT Press, Sandra Braman and Gita Devi Manaktala, as well as to Melinda Rankin, Michael Sims, and Kathy Caruso, who helped greatly improve the text. Mariel

García-Montes, Katie Louise Arthur, and Annis Rachel Sands each spent many hours working with me on the manuscript as research assistants and deserve special recognition. Maya Wagoner's thesis project greatly informed my thinking about design pedagogy. Also, the *#MoreThanCode* report and the T4SJ project, the findings of which are woven throughout this book, would never have happened without the work of Maya and Berhan Taye to plan, facilitate, and conduct interviews with key practitioners across the country. That project was also built through the hard work of Caroline Rivas and Chris Schweidler. Chris, thank you for all that you do to advance participatory action research. Sky House continues to be a sanctuary.

Special thanks to readers of early, often painfully dense draft chapters, especially Lilly Irani, Ruha Benjamin, Lisa Parks, Catherine D'Ignazio, Una Lee, and Alessandra Renzi. Justin Reich and Eric Klopfer provided invaluable comments on design pedagogy, and Cathy Hannabach and Summer McDonald at Ideas on Fire gave thoughtful early suggestions on the first chapter. Laura Forlano and all the organizers of the design feminisms track at Design Research Society 2018 created a space where I was able to openly discuss and further develop many of this book's themes. Lisha Nadkarni responded to the earliest draft of the book proposal with key suggestions. Casey Thoreson created the index.

This book would also never have happened without support from many of the faculty and staff at Comparative Media Studies/Writing (CMS/W) and across MIT, especially Lisa Parks, Jim Paradis, T. L. Taylor, Lara Baladi, Vivek Bald, Kat Cizek, Ian Condry, Karilyn Crockett, Paloma Duong, Fox Harrell, Eric Klopfer, Lorrie LeJeune, Ken Manning, Nick Montfort, Justin Reich, Ed Schiappa, William Uricchio, Jing Wang, Andrew Whittaker, Sarah Wolozin, and Ethan Zuckerman. The MIT Program in Women's and Gender Studies (WGS) has also been very kind to me, especially Helen Lee and Emily Hiestand, as has MIT CoLab, in particular Dayna Cunningham, Phil Thompson, and Ceasar McDowell. Others who provided great support and encouragement along the way include Arturo Escobar, Sadie Red Wing, Lina Dencik and the Data Justice project, and Joy Buolamwini.

Gracias a Shey Rivera, Dey Hernández, Luana Morales, Jasmine Gomez, Luis Cotto, y a todo el corillo de boricuas brillantes que da vida, amor y rabia en la lluvia con nieve de Boston.

Thank you to my mothers, Carol Chock (who provided a much-appreciated final round of review) and Barbara Zimbel, and my fathers, Peter Costanza and Paul Mazzarella. You've all been so wonderful and supportive over the years.

A Yara Liceaga Rojas: gracias en todas las dimensiones posibles e imposibles. También a Sol, Elías, e Inarú, semillas de un futuro mejor.

Series Editor's Introduction

Sandra Braman

It is a truism that in a democracy, every citizen should have the opportunity to take part in decision making about what we do, how we operate, how we structure the world in which we live. As theories, practices, and the organizational forms and processes of democracy developed over the past several hundred years, that was how we structured the social world within which we live. Elections, legislative processes, judicial interpretations of law and evidence have been and all remain parts of how we go about policy-making for humans.

Now, though, we recognize that the world for which we are making decisions is not just social. It is *sociotechnical*, and in the digital environment, as the saw says, "code is law," responsible for providing an infrastructure of constraints and affordances that affect what we do, how we operate, and how we structure the social world within which we live. The social side of decision making remains in place (however troubled), and it will be to everyone's advantage should we learn how to bring the two sides—the social and the technical—into a common conversation and decision making in concert. But what are the processes through which individuals might take part in the design of technologies hard and soft and of network architectures that are analogous to the processes we use to shape our social world? The better developed, the more sophisticated and nuanced, and the more widespread the practices and commitments of those who use technologies to participate in their creation, the more likely it is that communal efforts will effectively become a part of decision-making conversations and affect outcomes.

This is the problem to which Sasha Costanza-Chock's *Design Justice: Community-Led Practices to Build the Worlds We Need* devotes itself. Using illustrative cases involving a range of specific technologies, from airport body scanners to Twitter and more, the book moves systematically through the question of who it is who historically and typically participates in design processes—and who could; the sites, both physical and processual, where or during which communal participation in design processes can take place; and the limits to what even those individuals most avid about participating in such processes can do given all the pressures of their daily lives, on the one hand, and the multiple literacies needed on the other. The book is beautifully written, the cases inherently interesting in their own right, as well as importantly illustrative, and the author brings deep knowledge of several scholarly literatures to the work.

In all that it brings together, this book is a terrific contribution that should be widely read. University curricula that include degree programs in this domain are beginning to pop up all over; for these, *Design Justice* provides an exceptionally rich introductory overview of the field. Practitioners—both on the technology design side and on the social/political advocacy side—will find the work useful as well. But in my view the chapter that is the most interesting, original, and important is the third, on the ways in which social movements have themselves been responsible for much technological innovation. The other inadequately appreciated and studied social source of and motivation for technological invention and innovation, of course, is poverty; I thank the community of Guatemalan refugees in Minneapolis-St. Paul in the 1980s for teaching me this lesson so profoundly. There is literature in these two areas, but it is sporadic, often epiphenomenal. Christine Ogan, for example, is to be honored for research on the then-unexpected political uses of the videocassette recorder in Turkey, also in the 1980s, that was supported by grant funds received to study what were expected to be the Hollywood-intended purposes of increasing the distribution of films. The best of the work is strong in the thickness of case-driven analyses, frequently enriched by the personal experiences of authors who were participant-observers and/or who produced autoethnographies, but a full and systematic history of information and communication technologies from these interrelated perspectives has yet to be appear and would be incredibly valuable for many reasons.

And here the circle comes around. In today's sociotechnical environment, Costanza-Chock argues, the way to effectively ensure that there is design justice when it comes to technologies is to innovate on the social side with the development of new communal formations and processes. The book offers positive recommendations of multiple kinds, from practice to policy to research, throughout, and provides a good foundation from which to continue to develop and engage in research, theory, and praxis.

Preface

This book is about the relationship between design and power. It's about the growing community of designers, developers, technologists, scholars, educators, community organizers, and many others who are working to examine and transform design values, practices, narratives, sites, and pedagogies so that they don't continue to reinforce interlocking systems of structural inequality. It's about design, social justice, and the dynamics of domination and resistance at personal, community, and institutional levels. In essence, it's a call for us to heed the growing critiques of the ways that design (of images, objects, software, algorithms, sociotechnical systems, the built environment, indeed, everything we make) too often contributes to the reproduction of systemic oppression. Most of all, it is an invitation to build a better world, a world where many worlds fit; linked worlds of collective liberation and ecological sustainability.

Popular narratives of design, technology, and social change are dominated by techno-utopian hype about ever-more-powerful personal devices, "intelligent" systems, and "Twitter revolutions," on the one hand, and totalizing, pessimistic accounts of digital surveillance, disinformation, and algorithmic injustice, on the other. This book strives to ground our understanding of design, technology, and social change in the daily practices of activists and community organizers, who have always struggled to amplify the voices of their communities "by any media necessary."[1] As I hope to demonstrate, new information and communication technologies (ICTs) not only take shape in Silicon Valley, they also emerge from marginalized communities and social movement networks, both during waves of spectacular protest activity and also in everyday life. My broader goal is to advance the growing conversation about the pitfalls and possibilities of design as a tool for social transformation. I'll begin by sharing a story about my own embodied experience of trans* erasure, an experience that I believe contains valuable insights for nearly all design domains.

Introduction: #TravelingWhileTrans, Design Justice, and Escape from the Matrix of Domination

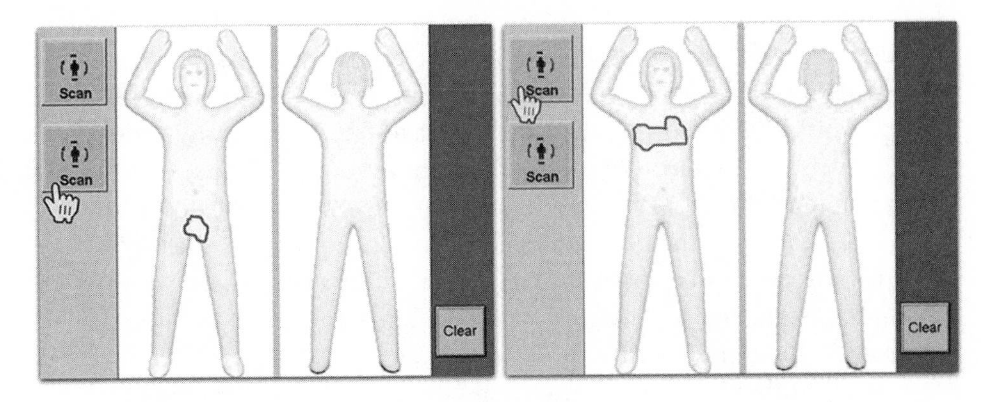

Figure 0.1
"Anomalies" highlighted in millimeter wave scanner interface. *Source:* Costello 2016.

It's June 2017, and I'm standing in the security line at the Detroit Metro Airport. I'm on my way back to Boston from the Allied Media Conference (AMC), a "collaborative laboratory of media-based organizing" that's been held every year in Detroit for the past two decades.[1] At the AMC, over two thousand people—media makers, designers, activists and organizers, software developers, artists, filmmakers, researchers, and all kinds of cultural workers—gather each June to share ideas and strategies for how to create a more just, creative, and collaborative world. As a nonbinary, trans*,[2] femme-presenting person, my time at the AMC was deeply liberating. It's a conference that strives harder than any that I know of to be inclusive of all kinds of people, including queer, trans*, intersex, and gender-non-conforming (QTI/GNC) folks. Although it's far from perfect, and every year inevitably brings new challenges and difficult conversations about what it means to construct a truly inclusive space, it's a powerful experience. Emerging from nearly a week immersed in this parallel world, I'm tired, but on a deep level, refreshed; my reservoir of belief in the possibility of creating better futures has been replenished.

Yet as I stand in the security line and draw closer to the millimeter wave scanning machine, my stress levels begin to rise. On one hand, I know that my white skin, US citizenship, and institutional affiliation with the Massachusetts Institute of Technology (MIT) place me in a position of relative privilege. I will certainly be spared the most disruptive and harmful possible outcomes of security screening. For example, I don't have to worry that this process will lead to my being placed in a detention center or in deportation proceedings; I won't be hooded and whisked away to Guantanamo Bay or to one of the many other secret prisons that form part of the global infrastructure of the so-called war on terror;[3] most likely, I won't even miss my flight while detained for what security expert Bruce Schneier describes as "security theater."[4] Only once in all of my travels have I been taken aside, placed into a waiting room, and subjected to additional questioning by the Department of Homeland Security (DHS).[5]

On the other hand, my heartbeat speeds up slightly as I near the end of the line, because I know that I'm almost certainly about to experience an embarrassing, uncomfortable, and perhaps humiliating search by a Transportation Security Administration (TSA) officer, after my body is flagged as anomalous by the millimeter wave scanner. I know that this is almost certainly about to happen because of the particular sociotechnical configuration of gender normativity (*cis-normativity*, or the assumption that all people have a gender identity that is consistent with the sex they were assigned at birth) that has been built into the scanner, through the combination of user interface (UI) design, scanning technology, binary-gendered body-shape data constructs, and risk detection algorithms, as well as the socialization, training, and experience of the TSA agents.[6]

A female-presenting TSA agent motions me to step into the millimeter wave scanner. I raise my arms and place my hands in a triangle shape, palms facing forward, above my head. The scanner spins around my body, and then the agent signals for me to step forward out of the machine and wait with my feet on the pad just past the scanner exit. I glance to the left, where a screen displays an abstracted outline of a human body. As I expected, bright fluorescent yellow pixels on the flat-panel display highlight my groin area (see figure 0.1). You see, when I entered the scanner, the TSA operator on the other side was prompted by the UI to select Male or Female; the button for Male is blue, the button for Female is pink. Since my gender presentation is nonbinary femme, usually the operator selects Female. However, the three-dimensional contours of my body, at millimeter resolution, differ from the statistical norm of female bodies as understood by the data set and risk algorithm designed by the manufacturer of the millimeter wave scanner (and its subcontractors), and as trained by a small army of clickworkers tasked with labeling and classification (as scholars Lilly Irani, Nick Dyer-Witheford, Mary Gray, and Siddharth Suri, among others, remind us).[7] If the agent selects Male, my breasts are large enough, statistically speaking, in comparison to the normative male body-shape construct in the database, to trigger an anomaly warning and a highlight around my chest area. If they select Female, my groin area deviates enough from the statistical female norm to trigger the risk alert. In other words, I can't win. This sociotechnical system is sure to mark me as "risky," and that will trigger an escalation to the next level in the TSA security protocol.

This is, in fact, what happens: I've been flagged. The screen shows a fluorescent yellow highlight around my groin. Next, the agent asks me to step aside, and (as usual) asks for my consent to a physical body search. Typically, once I'm close enough, the agent becomes confused about my gender. This presents a problem, because the next fork in the security protocol is for either a male or female TSA agent to conduct a body search by running their hands across my arms and armpits, chest, hips and legs, and inner thighs. According to TSA policy, "if a pat-down is performed, it will be conducted by an officer of the same gender as you present yourself."[8] As a nonbinary trans* femme, I present a problem not easily resolved by the algorithm of the security protocol. Sometimes, the agent will assume I prefer to be searched by a female agent; sometimes, a male. Occasionally, they ask for my preference. Unfortunately, "neither" is an honest but unacceptable response. Today, I'm particularly unlucky: a nearby male-presenting agent, observing the interaction, loudly states "I'll do it!" and strides over to me. I say, "Aren't you going to ask me what I prefer?" He pauses, then begins to move toward me again, but the female-presenting agent who is operating the scanner stops him. She asks me what I prefer. Now I'm standing in public, flanked by two TSA agents, with a line of curious travelers watching the whole interaction. Ultimately, the male-presenting agent backs off and the female-presenting agent searches me, making a face as if she's as uncomfortable as I am, and I'm cleared to continue on to my gate.

The point of this story is to provide a small but concrete example from my own daily lived experience of how larger systems—including norms, values, and assumptions—are encoded in and reproduced through the design of sociotechnical systems, or in political theorist Langdon Winner's famous words, how "artifacts have politics."[9] In this case, cis-normativity is enforced at multiple levels of a traveler's interaction with airport security systems. The database, models, and algorithms that assess deviance and risk are all binary and cis-normative. The male/female gender selector UI is binary and cis-normative.[10] The assignment of a male or female TSA agent to perform the additional, more invasive search is cis-normative and binary-gender normative as well. At each stage of this interaction, airport security technology, databases, algorithms, risk assessment, and practices are all designed based on the assumption that there are only two genders, and that gender presentation will conform with so-called biological sex. Anyone whose body doesn't fall within an acceptable range of "deviance" from a normative binary body type is flagged as risky and subjected to a heightened and disproportionate burden of the harms (both small and, potentially, large) of airport security systems and the violence of empire they instantiate. QTI/GNC people are thus disproportionately burdened by the design of millimeter wave scanning technology and the way that technology is used. The system is biased against us. Most cisgender people are unaware of the fact that the millimeter wave scanners operate according to a binary and cis-normative gender construct; most trans* people know, because it directly affects our lives.[11]

These systems are biased against QTI/GNC people, as I've described; against Black women, who frequently experience invasive searches of their hair, as documented by the team of investigative journalists at

ProPublica;[12] and against Sikh men, Muslim women, and others who wear headwraps, as described by sociologist Simone Browne in her brilliant book *Dark Matters*.[13] As Browne discusses, and as Joy Buolamwini, founder of the Algorithmic Justice League, technically demonstrates, gender itself is racialized: humans have trained our machines to categorize faces and bodies as male and female through lenses tinted by the optics of white supremacy.[14] Airport security is also systematically biased against Disabled people, who are more likely to be flagged as risky if they have non-normative body shapes and/or use prostheses, as well as anyone who uses a wearable or implanted medical device. Those who are simultaneously QTI/GNC, Black, Indigenous, people of color (PoC), Muslim, Sikh, immigrant, and/or Disabled[15] are doubly, triply, or multiply burdened by, and face the highest risk of harms from, this system.

I first publicly shared this experience in an essay for the *Journal of Design and Science* that I wrote in response to the "Resisting Reduction" manifesto, a timely call for thoughtful conversation about the limits and possibilities of artificial intelligence (AI).[16] That call resonated very deeply with me because as a nonbinary trans* feminine person, I walk through a world that has in many ways been designed to deny the possibility of my existence. The same cisnormative, racist, and ableist approach that is used to train the models of the millimeter wave scanners is now being used to develop AI in nearly every domain. From my standpoint, I worry that the current path of AI development will reproduce systems that erase those of us on the margins, whether intentionally or not, through the mundane and relentless repetition of reductive norms structured by the *matrix of domination* (a concept we'll return to later), in a thousand daily interactions with AI systems that, increasingly, weave the very fabric of our lives. My concerns about how the design of AI reproduces structural inequality extend more broadly to all areas of design, and these concerns are shared by a growing community.

The Design Justice Network

Design justice is not a term I created; rather, it emerged from a community of practice whose work I hope this book will lift up, extend, and support. This community is made up of design practitioners who participate in and work with social movements and community-based organizations (CBOs) across the United States and around the world. It includes designers, developers, technologists, journalists, community organizers, activists, researchers, and others, many of them loosely affiliated with the Design Justice Network (http://designjusticenetwork.org). The Design Justice Network was born at the AMC in the summer of 2015, when a group of thirty designers, artists, technologists, and community organizers took part in the workshop "Generating Shared Principles for Design Justice."[17] This workshop was planned by Una Lee, Jenny Lee, and Melissa Moore, and presented by Una Lee and Wesley Taylor. It was inspired by the Allied Media Projects (AMP) network principles, the Detroit Digital Justice Coalition (DDJC) digital justice principles, and the pedagogy of Detroit Future Youth. The goal of the workshop was to move beyond the frames of *social impact design* or *design for good*, to challenge designers to think about how good intentions are not necessarily enough to ensure that design processes and practices become tools for liberation, and to develop principles that might help design practitioners avoid the (often unwitting) reproduction of existing inequalities.[18] The draft principles developed at that workshop were refined by the Design Justice Network coordinators over the next year, revised at the AMC in 2017, and then, in 2018, released in the following form:

Design Justice Network Principles

This is a living document.

Design mediates so much of our realities and has tremendous impact on our lives, yet very few of us participate in design processes. In particular, the people who are most adversely affected by design decisions—about visual culture, new technologies, the planning of our communities, or the structure of our political and economic systems—tend to have the least influence on those decisions and how they are made.

Design justice rethinks design processes, centers people who are normally marginalized by design, and uses collaborative, creative practices to address the deepest challenges our communities face.

1. We use design to **sustain, heal, and empower** our communities, as well as to seek liberation from exploitative and oppressive systems.
2. We **center the voices of those who are directly impacted** by the outcomes of the design process.
3. We **prioritize design's impact on the community** over the intentions of the designer.

4. We view **change as emergent from an accountable, accessible, and collaborative process**, rather than as a point at the end of a process.
5. We see the role of the **designer as a facilitator rather than an expert.**
6. We believe that **everyone is an expert based on their own lived experience**, and that we all have unique and brilliant contributions to bring to a design process.
7. We **share design knowledge and tools** with our communities.
8. We work towards **sustainable, community-led and controlled** outcomes.
9. We work towards **non-exploitative solutions** that reconnect us to the earth and to each other.
10. Before seeking new design solutions, **we look for what is already working** at the community level. We honor and uplift traditional, indigenous, and local knowledge and practices.[19]

These principles have now been adopted by over three hundred people and organizations. The Design Justice Network has grown, nurtured by many; besides dozens of track coordinators (many named in this book's acknowledgments) and workshop facilitators, ongoing steering committee members include designers Una Lee, Victoria Barnett, Wesley Taylor, and myself.[20] The network produces a series of zines that provide an evolving record of our ideas and activities (http://designjusticenetwork.org/zine); coordinates a track at the AMC; and organizes workshops on a regular basis. Information about the dozens of organizations and hundreds of individuals that have been part of the design justice track at AMC is available in the archived conference programs.[21]

In particular, the design studio And Also Too has been a key actor in the development of design justice ideas and practices. Founded by designer Una Lee, And Also Too is "a collaborative design studio for social justice visionaries," and is home to designers and artists Lupe Pérez, Sylver Sterling, Lara Stefanovich-Thomson, and Zahra Agjee. As they describe on their site: "And Also Too uses co-design to create tools for liberation and visionary images of the world we want to live in. … Our work is guided by two core beliefs: first, that those who are directly affected by the issues a project aims to address must be at the center of the design process, and second, that absolutely anyone can participate meaningfully in design."[22] And Also Too facilitated the development of the Design Justice Network Principles, and is guided by those principles in its own day-to-day work.[23] Others that practice design justice include the worker-owned cooperative Research Action Design (RAD),[24] the Detroit-based artist collective Complex Movements, and a growing list of more than three hundred Design Justice Network Principles signatories (the full list is available at http://designjusticenetwork.org/network-principles).

More recently, other groups that are not (yet!) formally connected to the Design Justice Network have also begun to use the hashtag #designjustice on various social media platforms. These include the architects and city planners who organized a series of DesignAsProtest events in 2017, the EquityXDesign campaign to end gender and racial disparity in architecture as a profession, and the architects affiliated with the American Institute of Architects (AIA) who convened a 2018 Design Justice Summit in New Orleans, among others. The Equity Design Collaborative, led by Caroline Hill, Michelle Molitor, and Christine Ortiz, has been working to retrofit design thinking methods with a racial justice analysis.[25]

There are also many, many organizations that don't use the term *design justice* but are engaged in closely allied practices. For example, the Inclusive Design Research Centre (IDRC) is "a research and development centre where an international community of open source developers, designers, researchers, advocates, and volunteers work together to ensure that emerging information technology and practices are designed inclusively."[26] Professor of Civic Design Ceasar McDowell has developed an extensive body of theory and practice of *design for the margins*.[27] Other allied projects, groups, and networks include the Association for Progressive Communications, the Catalan GynePunk collective (who develop and circulate queer feminist design practices of DIY gynecology[28]), the Center for Media Justice, Coding Rights (Brazil), the Critical Making Lab, Data Active, Decolonising Design, the Design Studio for Social Intervention, Design Trust for Public Space, the Digital Justice Lab (Toronto), FemTechNet, Intelligent Mischief (Brooklyn), MIT CoLab, SEED Network, Social Justice Design Studio, and the Tech Equity Collective, just to name a few.[29]

In particular, there is a rapidly growing community of researchers, computer scientists, and advocates who are focused on challenging the ways that inequality is reproduced through the design of AI and algorithmic decision support systems. This area has seen a wave of recent publications, such as Virginia Eubanks's *Automating Inequality* (2018), Safiya Noble's *Algorithms of Oppression* (2018), Meredith Broussard's *Artificial Unintelligence* (2019), and Ruha Benjamin's *Race After Technology* (2019), among others. In this area, there is also an explosion of new organizations and networks. Data for Black Lives has emerged as a key community of data scientists, scholars, artists, and community organizers who work to rethink data science, machine learning, AI, and other sociotechnical systems through a racial justice lens. Others (among many!) include the AI Now Institute, the Algorithmic Justice League, the Center for Critical Race and Digital Studies, Data & Society, the Data Justice Lab (Cardiff), the Digital Equity Lab (NYC), the JUST DATA Lab, the Our Data Bodies Project, the People's Guide to AI, and the Stop LAPD Spying Coalition.

Throughout this book, I will return to, draw from, and reference the work of these and other scholars, designers, and organizations that are already working to put design justice principles into practice, although there are so many that it won't be possible to mention them all.

Methods

My Own Standpoint

Feminist standpoint theory recognizes that all knowledge is situated in the particular embodied experiences of the knower.[30] Accordingly, I begin here by locating my own position and trajectory for the reader. I'm a nonbinary trans* femme queer person, of Italian-Russian-Polish-Jewish descent, raced white within the current logic of racial capitalism in the United States. I was born into a rural, hippie, cooperative home near Ithaca, in upstate New York, to parents who took part in feminist, antiwar, anti-imperialist, Latin American solidarity, and environmentalist movements of the time. I grew up on land stolen from the Onöñda'gaga' (Onandaga), Susquehannock, Gayogohó:no' (Cayuga), and peoples of the Haudenosaunee (Iroquois) confederacy. My political education came first via my parents and community, then my teachers at the Alternative Community School, a public alternative school. I attended high school in Puebla, México, then moved to Boston and attended Harvard College on a scholarship, gaining access to a new level of educational privilege. While in Boston, I joined the popular theater and cultural organizing collective AgitArte,[31] and in that work became more deeply politicized through the efforts of Puerto Rican artist-organizers like Jose Jorge Díaz and Mayda Grano de Oro. After college, I lived and worked in San Juan, Puerto Rico, with the public arts project EducArte, before moving to Philadelphia for graduate education, hoping to connect my activist work to media theory.

At that time, in the early 2000s, I was part of the global Indymedia network of DIY social movement journalism.[32] I traveled throughout Latin America to bring donated video cameras and computers to local Indymedia collectives, participated in organizing Independent Media Centers to provide grassroots coverage of large protest events, and produced and distributed documentary films and videos about the global justice movement.[33] Through Indymedia, I also learned about free software and gained software development skills.

In 2003, I became involved with the Allied Media Conference, a space that continues to transform and shape my life.[34] I moved to Los Angeles for a PhD program at the University of Southern California, and while there, I worked with the Institute of Popular Education of Southern California (IDEPSCA), the Garment Worker Center (GWC), and other community-based organizations to support worker-led media projects like VozMob (*Voces Móviles*/Mobile Voices), developed through participatory design.[35] In 2011, I moved to Boston to take a position at MIT, and in 2014, I cofounded the worker-owned cooperative Research Action Design with Chris Schweidler and Bex Hurwitz.

As I write these words, in 2018, I have a faculty position at a high-profile university. I materially benefit from, and in some ways am harmed by, my location within systems including whiteness, educational inequality, capitalism, ableism, and settler colonialism. Simultaneously, I experience oppression based on patriarchy (although in the past I experienced both benefits and harms from this system), transphobia, transmisogyny, and cis-normativity. My standpoint and lived experience shape my understanding of design as a tool for both oppression and liberation, and throughout this text I will occasionally return to my lived experience to ground and illustrate key points.

Participatory Action Research

Most of my work falls within the tradition of participatory action research (PAR) and codesign. PAR is a framework with roots in the work of scholars and educators such as Kurt Lewin, John Dewey, and (later) Paulo Freire, Orlando Fals-Borda, and Linda Tuhiwai Smith, and it emphasizes the development of communities of shared inquiry and action.[36] *Codesign*, a closely allied approach, can be traced to Scandinavian efforts in the 1960s and 1970s to include both workers and managers in sociotechnical systems design. Both PAR and codesign consider communities to be co-researchers and co-designers, rather than solely research subjects or test users. Chapter 2 provides a more in-depth discussion of the roots of codesign methods.

Together with the community-based organizations that are my research partners, I typically employ a combination of participant observation, semi-structured interviews, popular education, and codesign workshops. The empirical grounding for this book includes (1) my experience as a cofounder of Research Action Design (RAD.cat), a worker-owned cooperative that attempts to put the principles of design justice into action; (2) my work as part of the Tech for Social Justice Project, a PAR team that produced the report *#MoreThanCode: Practitioners Reimagine the Landscape of Tech for Justice and Equity*,[37] based on more than one hundred semi structured interviews (most of them conducted by Maya Wagoner and Berhan Taye) and a series of eleven focus groups with technologists, designers, developers, product managers, and others across the United States (explore morethancode.cc); and (3) my own experience developing, teaching, and evaluating the Civic Media: Collaborative Design Studio course at MIT, from 2012 through the present (https://codesign.mit.edu).

Thus, although this book itself is not a PAR project, the experiences and insights that it contains were developed over many years in community and in collaboration with other researchers, community organizers, and design practitioners.

A Note on "We" and "I"

As an engaged scholar and design practitioner who is guided by antiracist, feminist principles and epistemology, I want to make clear that although this is a single-authored book, many of the ideas it explores have bubbled up through the Design Justice Network as an emergent community of practice. All credit for the key ideas of design justice is due to this community, whereas all responsibility for the many errors in this text is mine. To paraphrase one of the anonymous reviewers of this manuscript, there is a tension between my attempt to provide a normative design justice framework as a single author and my claim to be amplifying knowledge that has been produced by a movement. I will do my best to remind the reader of this tension throughout.

In this book, I also move back and forth between third-person description and use of the first-person pronouns *we* and *I*. In particular, I use the first-person singular when I'm describing or drawing from my own personal experience to illustrate a point. When I use *we*, sometimes it refers to the community of existing design justice practitioners, and I will attempt to make that clear. At other moments, *we* refers to the aspirational broader community of those who care about remaking design, as part of broader efforts to make more liberatory and just worlds. I hope that you (the reader) will feel included in this broader *we*. Let us begin with a few key terms.

Design Justice: Defining Key Terms

Design

> Design (noun): A plan or scheme conceived in the mind and intended for subsequent execution; the preliminary conception of an idea that is to be carried into effect by action; a project.
>
> —*Oxford English Dictionary*[38]

There are many definitions of *design*. I won't attempt their synthesis here, nor will I advocate for the adoption of a particular definition. Nevertheless, before diving into the theory and practice of design justice, I'll briefly discuss a few of the many ways that the term *design* is used and offer some thoughts about the meanings that are most useful in the context of this book.

As a verb, *design* originates from the Latin *de signum* ("to mark out") or *designō* ("I mark out, point out, describe.") In early use, it described the act of making a meaningful physical mark on an object. *Signum* evolved, mostly through French, into words such as "signify, assign, designate, [and] signal,"[39] and this sense is maintained today in the idea that designers sketch, draw, and mark out representations that will later become objects, buildings, or systems. In common usage, *design* carries multiple meanings. We use it to refer to a plan for an artifact, building, or system; a pattern (such as a floral print on a textile); the composition of a work of art; or the shape, appearance, or features of an object.[40] It also refers to the practice, field, or subfields of design work (e.g., "Icelandic design dominates global furniture markets.").

In his classic text *Design for the Real World,* Victor Papanek positions design as a universal practice in human communities: "All [people][41] are designers. ... Design is the conscious effort to impose a meaningful order."[42] Design professor, practitioner, and philosopher Tony Fry also argues that we are all designers and that design is not solely the province of architects, graphic designers, industrial designers, or other design professionals; instead, he sees it as a component of all intentional acts.[43] Anne-Marie Willis, professor of design theory and editor of *Design Philosophy Papers,* puts it this way:

> Design is something far more pervasive and profound than is generally recognised by designers, cultural theorists, philosophers or lay persons; designing is fundamental to being human—we design, that is to say, we deliberate, plan and scheme in ways which prefigure our actions and makings ... we design our world, while our world acts back on us and designs us.[44]

At the same time, *design* frequently refers to expert knowledge and practices contained within a particular set of professionalized fields, including graphic design, fashion design, interaction design, industrial design, architecture, planning, and various other industries. Alongside the discussion of design as a specialist activity or as a certain type of work accomplished by experts, there is also a steadily growing literature on marginalized people's design practices. In line with feminist critiques of frequently unpaid and invisibilized forms of feminized labor,[45] it's crucial to acknowledge the importance of everyday, vernacular, and often unrecognized design practices (as in chapter 3). Alternative histories of technology and design help to recuperate and center people, practices, and forms of expertise that have long been erased by mainstream design theory and history, both in scholarly and popular writing. A few of these counter histories of invisibilized technology design work have been widely popularized; for example, the 2016 film *Hidden Figures* chronicles the work of Katherine Johnson and other Black women who worked for NASA as "human computers," coding space flight trajectories.[46] In addition, recent innovation literature decenters the myth of the individual designer and emphasizes the key roles played by "lead users" who constantly modify, hack, repurpose, and reuse technologies to better fit their needs[47] (a point taken up in chapter 2).

However, inclusive visions of design as a universal human activity in many ways conflict with the realities of the political economy of design. True, everyone designs, but only certain kinds of design work are acknowledged, valorized, remunerated, and credited. In other words, design is professionalized: certain people get paid, sometimes quite well, to be design experts. Designers have professional associations (such as the American Institute of Graphic Arts, or AIGA, with over twenty-five thousand members),[48] conferences, and in some subfields, extensive processes for accreditation and licensing (architects, industrial designers), standardization (negotiated through standards bodies such as the United States Access Board, tasked with developing the Americans with Disabilities Act Accessibility Guidelines), norms, and principles (such as universal design principles).[49]

According to design scholars Robert Hoffman, Axel Roesler, and Brian Moon, the *designer* as a specific kind of person, or as a profession, emerged with the Industrial Revolution. Until then, knowledge about how to create, use, and maintain specialized tools was transmitted via craft guilds. However, the craft guild model could not support larger-scale designs that required the distribution of skills among many specialists. Accordingly, "this new task—designing for a class of people with whom the designer did not interact—helped mark the origin of industrial design."[50] At this time, they also note, designers took on a new role: "to reshape formerly hand-crafted processes into ones that machines could do. Mass and assembly-line-based production stimulated, or necessitated, the creation of many designs for artifacts aimed at a broad mass of consumers and for machines designed to help in manufacturing other machines."[51]

The Industrial Revolution–era association of design with industry, machines, and mass production shifted over time. Design, designers, and design work are now inextricably linked with computers, software, and the virtual representation of objects and systems. Across all professional design fields, including industrial design, architecture, graphic design, and software design, design work has become primarily digital work, performed with computers and software tools. As in so many fields, certain design tasks are also increasingly automated or semiautomated. In chapter 2, I will further discuss the implications of design justice on the question of who gets paid to do design work.

Design is also a way of thinking, learning, and engaging with the world. Reasoning through design is a mode of knowledge production that is neither primarily deductive nor inductive, but rather abductive and speculative. Where *deduction* reasons from the general to the specific and *induction* reasons from the specific to the general, *abduction* suggests the best prediction given incomplete observations.[52] Professor of urban planning, philosopher, and scholar of organizational learning Donald Schön put it this way: "Designers put things together and bring new things into being, dealing in the process with many variables and constraints, some initially known and some discovered through designing. Almost always, designers' moves have consequences other than those intended for them. Designers juggle variables, reconcile conflicting values, and maneuver around constraints—a process where, although some design products may be superior to others, there are no unique right answers."[53] Design is thus also *speculative*: it is about envisioning, as well as manipulating, the future.[54] Designers imagine images, objects, buildings, and systems that do not yet exist. We propose, predict, and advocate for (or, in certain kinds of design, warn against) visions of the future.

In his recent book *Designs for the Pluriverse* (2018), anthropologist Arturo Escobar sees design as an "ethical praxis of world-making."[55] He urges us to consider the ways that design practices today too often reproduce the totalizing epistemology of modernity and in the process erase indigenous worldviews, forms of knowledge, and ways of being. Escobar calls for an approach to design that is focused on the creation of a world "where many worlds fit." This is a reference to the Zapatista slogan that so powerfully articulates a need to move past the current globalized system that is spiraling rapidly toward ecological collapse. Escobar reminds us that the erasure of indigenous lifeworlds takes place through the long-running and still-unfolding imposition of colonial ontologies, epistemologies, and ways of knowing the world. The call for community-led practices to build the *worlds* we need (this book's subtitle) is directly inspired by Escobar's discussion of the pluriverse. In a similar vein, Ramesh Srinivasan, in his recent book *Whose Global Village?*, reminds us that indigenous peoples have their own ways of imposing meaningful order on the world, which have not only been under attack through centuries of colonialism but also are often erased in interactions with present-day sociotechnical systems, even within supposedly human-centered or participatory design processes.[56]

What of design itself as a totalizing project? Undoubtedly, *design thinking* has become increasingly popular. Propelled by the Stanford d.school and by the design firm IDEO, this approach is widely influential throughout business, the academy, and, most recently, the public sector.[57] Feminist science and technology studies (STS), human-computer interaction (HCI), and South Asia studies scholar Lilly Irani critiques the way that design thinking is deployed to reproduce a colonial political economy, with design imagined at the top of the value chain as a key process to be managed only by firms from the Global North (and as a mechanism for the reproduction of whiteness).[58] Product designer Natasha Jen, in a widely seen 99U talk, states that "design thinking is bullshit."[59] Sociologist Ruha Benjamin, in her recent book *Race After Technology: Abolitionist Tools for the New Jim Code* (2019), examines the relationship between design and systemic racism; she calls both for a more intentionally antiracist approach to innovation and for a healthy skepticism of universalist and solutionist notions of design as a way out of structural inequality.[60] I will return to a discussion of design thinking later in the book.

Design thus may be thought of as both a verb and a noun, a universal kind of human activity and a highly professionalized field of practice (or several such fields), a way of manipulating future objects and systems using specialized software and an everyday use of traditional knowledge embedded in indigenous lifeways, a type of work with one's hands and a way of thinking, an art and a science, and more. My goal is not to capture or reduce this multivalence to a single true essence. Instead, design justice raises a set of questions and provocations that (I believe) apply to *any and all* meanings of design. Before I offer a working definition of *design justice*, however, I will briefly discuss two key concepts from Black feminist thought that reside at the core of many of this book's arguments: *intersectionality* and the *matrix of domination.*

Intersectionality

Black feminist thought fundamentally reconceptualizes race, class, and gender as interlocking systems: they do not only operate on their own, but are often experienced together by individuals who exist at their intersections. The analytical framework built on this fundamental insight is called *intersectionality*. Although the idea has a longer legacy (think of African American abolitionist and women's rights activist Sojourner Truth's "Ain't I a Woman?," Communist Party Secretary Claudia Jones's writings about being "triply oppressed," or the Combahee River Collective's critiques of white feminism),[61] the specific term *intersectionality* was first published by Black feminist legal scholar Kimberlé Crenshaw in her 1989 article "Demarginalizing the Intersection of Race and Sex: A Black Feminist Critique of Antidiscrimination Doctrine, Feminist Theory and Antiracist Politics."[62] In the article, Crenshaw describes how existing antidiscrimination law (Title VII of the Civil Rights Act) repeatedly failed to protect Black women workers.

First, she discusses an instance in which Black women workers at General Motors (GM) were told they had no legal grounds for a discrimination case against their employer because antidiscrimination law only protected single-identity categories. The court found that, since GM hired white women, the company did not systematically discriminate against women. It further found that there was insufficient evidence of discrimination against Black people, because GM hired significant numbers of Black men to work on the line. Thus, Black women, who in reality did experience systematic employment discrimination *as Black women,* were not protected by existing law and had no actionable legal claim. In a second case described by Crenshaw, the court rejected the discrimination claims of a Black woman who sued Hugh Helicopters, Inc., because "her attempt to specify her race was seen as being at odds with the standard allegation that the employer simply discriminated 'against females.'"[63] In other words, the court could not accept that Black women might be able to represent *all* women, including white women, as a class. In a third case, the court *did* award discrimination damages to Black women workers at a pharmaceutical company, but it refused to award the damages to *all* Black workers, under the rationale that Black women could not possibly represent the claims of Black people as a whole.[64]

Crenshaw notes the role of statistical analysis in each of these cases: sometimes, the courts required Black women plaintiffs to include broader statistics for all women that countered their discrimination claims; in other cases, the courts limited the admissible data to that which dealt solely with Black women, as opposed to all Black workers. In those cases, the low total number of Black women employees typically made statistically valid discrimination claims impossible, whereas strong claims could have been made if the plaintiffs were allowed to include data for all women, for all Black people, or both. Later, in her 1991 *Stanford Law Review* article "Mapping the Margins: Intersectionality, Identity Politics, and Violence against Women of Color,"[65] Crenshaw powerfully articulates the ways that women of color often experience male violence as a product of intersecting racism and sexism, but are then marginalized from both feminist and antiracist discourse and practice and denied access to specific legal remedies.[66]

The concept of intersectionality provided the grounds for a long, slow paradigm shift that is still unfolding in the social sciences, in legal scholarship, and in other domains of research and practice. This paradigm shift is also beginning to transform the various domains of design. One of the central claims of this book is that the predominance of what Crenshaw calls *single-axis analysis*, in which race, class, or gender is considered as an independent construct, continually undermines the intentions of well-meaning designers who hope to challenge bias through the objects, systems, or environments they design. In law, as Crenshaw points out, "the single-axis framework erases Black women in the conceptualization, identification and remediation of race and sex discrimination by limiting inquiry to the experiences of otherwise-privileged members of the group. In other words, in race discrimination cases, discrimination tends to be viewed in terms of sex- or class-privileged Blacks; in sex discrimination cases, the focus is on race- and class-privileged women. This focus on the most privileged group members marginalizes those who are multiply-burdened and obscures claims that cannot be understood as resulting from discrete sources of discrimination."[67]

In this book, I will demonstrate how universalist design principles and practices erase certain groups of people, specifically those who are intersectionally disadvantaged or multiply burdened under white supremacist heteropatriarchy, capitalism, and settler colonialism. What is more, when designers do consider inequality in design (and most professional design processes do not consider inequality at all), they nearly always employ a single-axis framework. Most design processes today therefore are structured in ways that make it impossible to see, engage with, account for, or attempt to remedy the unequal distribution of benefits and burdens that they reproduce. As Crenshaw notes, feminist theory and antiracist policy that is not grounded in an intersectional understanding of gender, race, and class can never adequately address the experiences of Black women, or any other multiply burdened groups of people, when it comes to the formulation of policy demands. Design justice holds that the same is true when it comes to "design demands."

For example, intersectionality is an absolutely crucial concept for the development of AI. Most pragmatically, single-axis (in other words, nonintersectional) algorithmic bias audits are insufficient to ensure algorithmic fairness (let alone justice). While there is rapidly growing interest in algorithmic bias audits, especially in the fairness, accountability, and transparency in machine learning (FAT*) community, most are single-axis: they look for a biased distribution of error rates only according to a single variable, such as race or gender. This is an important advance, but it is essential that we develop a new norm of intersectional bias audits for machine learning systems. Toward that end, Joy Buolamwini of the Algorithmic Justice League has produced a growing body of work that demonstrates the ways that machine learning is intersectionally biased. In the project *Gender Shades*, Buolamwini and researcher Timnit Gebru show how facial analysis tools trained on "pale male" data sets perform best on images of white men and worst on images of Black women.[68] In order to demonstrate this, they first had to create a new benchmark data set of images of faces, both male and female, with a range of skin tones.

Of course, there are many cases where a design justice analysis asks us not to make systems more inclusive, but to refuse to design them at all; we will return to that point repeatedly as well as at the end of the book in a discussion of the #TechWontBuildIt movement. However, industry appropriation aside, Buolamwini and Gebru's work not only demonstrates that facial analysis systems are technically biased (although that is true); it also provides a concrete example of the lesson that, wherever we contemplate developing machine learning systems, we need to develop intersectional training data sets, intersectional benchmarks, and intersectional audits. The urgency of doing so is directly proportional to the impacts (or potential impacts) of algorithmic decision support systems on people's life chances.

More broadly, without intersectional analysis, we cannot design any objects or systems that adequately address the experiences of people who are multiply burdened within the matrix of domination.

The Matrix of Domination

Closely linked to intersectionality, but less widely used today, the *matrix of domination* is a term developed by Black feminist scholar, sociologist, and past president of the American Sociological Association Patricia Hill Collins to refer to race, class, and gender as interlocking systems of oppression. It is a conceptual model that helps us think about how power, oppression, resistance, privilege, penalties, benefits, and harms are systematically distributed. When she introduces the term in her 1990 book *Black Feminist Thought,* Collins emphasizes race, class, and gender as the three systems that historically have been most important in structuring most Black women's lives. She notes that additional systems of oppression structure the matrix of domination for other kinds of people. The term, for her, describes a mode of analysis that includes any and all systems of oppression that mutually constitute each other and shape people's lives.[69]

Collins also emphasizes that every individual simultaneously receives both benefits and harms based on their location within the matrix of domination. As Collins notes, "Each individual derives varying amounts of penalty and privilege from the multiple systems of oppression which frame everyone's lives."[70] An intersectional Black feminist analysis thus helps us each understand that we are simultaneously members of multiple dominant groups and multiple subordinate groups. Design justice urges us to (1) consider how design (affordances and disaffordances, objects and environments, services, systems, and processes) distributes both penalty and privileges to individuals based on their location within the matrix of domination and (2) attend to the ways that this operates at various scales.

In *Black Feminist Thought*, Collins also notes that "people experience and resist oppression on three levels: the level of personal biography; the group or community level of the cultural context created by race, class, and gender; and the systemic level of social institutions. Black feminist thought emphasizes all three levels as sites of domination and as potential sites of resistance."[71] Design justice urges us to explore the ways that design relates to domination and resistance at each of these three levels (personal, community, and institutional). For example, at the personal level, we might explore how interface design affirms or denies a person's identity through features such as a binary gender dropdown menu during profile creation. Such seemingly small design decisions have disparate impacts on different individuals.

At the community level, platform design (for example) fosters certain kinds of communities while suppressing others, through setting and enforcing community guidelines, rules, and speech norms, instantiated through different kinds of content-moderation algorithms, clickworkers, and decision support systems. For example, when ProPublica revealed that Facebook's internal content moderation guidelines explicitly mention that Black children are not a protected category, while white men are,[72] this inspired very little confidence in Mark Zuckerberg's congressional testimony that Facebook feels it can deal with hate speech and trolls through the use of AI content moderation systems. Nor was Facebook's position improved by the leak of content moderation guidelines that note that "white supremacist" posts should be banned, but that "white nationalist" posts are within free speech bounds.[73]

At the institutional level, we might consider how design decisions that reproduce and/or challenge the matrix of domination are influenced by institutional funding priorities, policies, and practices. Design institutions include companies (Google, Apple, Microsoft), nation-states that decide what kinds of design to prioritize through funding agencies such as the National Science Foundation (NSF) and Department of Defense (DoD), venture capital firms, standards-setting bodies (like ISO, W3C, and NIST), laws (such as the Americans with

Disabilities Act), universities that educate designers, and so on. Not only do institutions influence design by other actors, they also design objects, systems, and processes that they then use to distribute benefits and harms across society. For example, the ability to immigrate to the United States is unequally distributed among different groups of people through a combination of laws passed by the US Congress, software decision support systems, executive orders that influence enforcement priorities, and so on. In 2018, the Department of Homeland Security had an open bid to develop "extreme vetting" software that would automate "good immigrant/bad immigrant" prediction by drawing from people's public social media profiles. After extensive pushback from civil liberties and immigrant rights advocates, DHS backpedaled and stated that the system was beyond "present-day capabilities." Instead, they announced a shift in the contract from software to labor: more than $100 million dollars will be awarded to cover the employment of 180 people, tasked with manually monitoring immigrant social media profiles from a list of about one hundred thousand people.[74] More broadly, visa allocation has always been an algorithm, one designed according to the political priorities of power holders. It's an algorithm that has long privileged whiteness, hetero- and cis-normativity, wealth, and higher socioeconomic status.[75]

Finally, Black feminist thought emphasizes the value of situated knowledge over universalist knowledge. In other words, particular insights about the nature of power, oppression, and resistance come from those who occupy subjugated standpoints. This approach also explicitly recognizes that knowledge developed from any particular standpoint is partial knowledge: "The overarching matrix of domination houses multiple groups, each with varying experiences with penalty and privilege that produce corresponding partial perspectives, situated knowledges, and, for clearly identifiable subordinate groups, subjugated knowledges. No one group has a clear angle of vision. No one group possesses the theory or methodology that allows it to discover the absolute 'truth' or, worse yet, proclaim its theories and methodologies as the universal norm evaluating other groups' experiences."[76]

The challenges presented by deeply rooted and interlocking systems of oppression can seem overwhelming. What paths might lead us out of the matrix of domination?

Design Justice

So far, we have briefly explored the meanings of *design*, *intersectionality*, and the *matrix of domination*. To conclude this section, I offer the following tentative description of design justice:

> Design justice is a framework for analysis of how design distributes benefits and burdens between various groups of people. Design justice focuses explicitly on the ways that design reproduces and/or challenges the matrix of domination (white supremacy, heteropatriarchy, capitalism, ableism, settler colonialism, and other forms of structural inequality). Design justice is also a growing community of practice that aims to ensure a more equitable distribution of design's benefits and burdens; meaningful participation in design decisions; and recognition of community-based, Indigenous, and diasporic design traditions, knowledge, and practices.

This isn't meant to be a canonical definition of design justice. Nor should it supplant the Design Justice Network Principles presented earlier, which were developed by a growing community of practitioners through an extensive, multiyear process. Instead, it is a provisional, succinct description that I found useful as I worked to organize my thoughts about design theory and practice for this book.

This description of design justice also resonates strongly with the current widespread rise of intersectional feminist thought and action, visible in recent years in the United States in the emergence of networked social movements such as #BlackLivesMatter, the immigrant rights movement, the fight for LGBTQI+ and Two-Spirit rights, gender justice, and trans* liberation, indigenous struggles such as #IdleNoMore and #StandWithStandingRock, disability justice work, the #MeToo movement, the environmental justice movement, and new formations in the labor movement such as platform cooperativism and #TechWontBuildIt. These movements fight to resist the resurgent extreme right, and also to advance concrete proposals for a more just and sustainable world. They are growing, and in 2018 provided the momentum for a historic midterm election that won record numbers of seats for leftists, queer people, and B/I/PoC in the US Congress.

Intersectional feminist networked movements are also increasingly engaged in debates about the relationships between technology, design, and social justice. It is my hope that design justice as a framework can provide tools to support existing and emergent critique of design (from images to institutions, from products to platforms, from particular practitioners to professional associations), as well as encourage the documentation of innovative forms of community-led design, grounded in the specificity of particular social movements. In this book, I draw from the activities of the Design Justice Network, my own experience working on design projects and teaching design theory and practice, practitioner interviews, and texts by other scholars, designers, and community organizers. I hope that this book can help shift our conversation beyond the need for diversity in tech-sector employment, and that it will help make visible the growing community of design justice practitioners who are already working closely with liberatory social movements to build better futures for us all.

Chapter Overview

In the Design Justice Network, for the last several years we have been asking questions about how design currently works, and about how we want it to work. I have structured the chapters in this book as an extensive reflection on a few of these questions—in particular:

- *Values.* What values do we encode and reproduce in the objects and systems that we design?
- *Practices.* Who gets to do design? How do we move toward community control of design processes and practices?
- *Narratives.* What stories do we tell about how things are designed? How do we scope design challenges and frame design problems?
- *Sites.* Where do we do design? How do we make design sites accessible to those who will be most impacted by design processes? What design sites are privileged and what sites are ignored or marginalized?
- *Pedagogies.* How do we teach and learn about design justice?

The book is organized as follows:

Chapter 1 addresses the question, "What values do we encode and reproduce in the objects and systems that we design?" It argues that, currently, the values of white supremacist heteropatriarchy, capitalism, ableism, and settler colonialism are too often reproduced in the affordances and disaffordances of the objects, processes, and systems that we design. The chapter begins with a story about using Facebook to organize a trans*, queer, and immigrant solidarity protest and uses that experience to open a critical conversation with the literature on affordances, disaffordances, discriminatory design, and cognitive load.[77] Although design affordances are often assumed to be universal, the chapter argues that they are actually unequally distributed based on the matrix of domination. The next section briefly discusses approaches such as value-sensitive design,[78] universal design, and inclusive design. Over time, these have produced much-needed shifts in design theory and practice, and design justice builds upon them but also differs in important ways. The chapter also draws on feminist and antiracist strands within science and technology studies to unpack the ways that the matrix of domination is constantly hard-coded into designed objects and systems.[79] This typically takes place not because designers are intentionally "malicious" but through unintentional mechanisms, including assumptions about "unmarked" end users, the use of systematically biased data sets to train algorithms using machine-learning techniques, and limited feedback loops. Addressing these issues requires that we retool for design justice, and the chapter analyzes various design concepts and tools, such as differential cognitive load, intersectional instrumentation, benchmarking, and A/B testing, through a design justice lens. It ends with a question about what it might mean to hard-code liberation.

Chapter 2 focuses on the questions, "Who gets to do design? How do we move toward community control of design processes and practices?" It argues that the most valuable ingredient in design justice is the full inclusion of, accountability to, and control by people with direct lived experience of the conditions designers claim they are trying to change. The chapter builds on the work of the disability justice movement, whose activists popularized the phrase "nothing about us without us."[80] It begins with a discussion of the raced, classed, and gendered nature of employment in the technology sector, but quickly proposes a shift from arguments for equity (such as "we need more diverse designers and software developers") to arguments for accountability and community control ("those most affected by the outcomes should lead and own design processes and products"). This is not a new idea; the chapter reviews the participatory turn in technology design and includes discussion of user-led innovation, participatory design, and feminist HCI, among other strands of theory and practice.[81] Key lessons include the following: leadership and control by members of the community that is most directly affected by the issue is crucial, both because it's ethical and also because the tacit and experiential knowledge of those marginalized within the matrix of domination is sure to produce ideas, approaches, and innovations that a nonmember of the community would be very unlikely to come up with. The chapter ends by exploring findings from the #MoreThanCode field scan of technology for social justice practitioners across the United States; in particular, it summarizes practitioners' suggestions about how to create community accountability in technology design processes.

Stories have great power, and chapter 3 asks, "What stories do we tell about the design of digital technologies?" It opens by contrasting the "official" Twitter origin story (one of the founders had a brilliant blue-sky flash of genius) with counternarratives from developers who were part of the process (anarchist activists created the demo design for Twitter as a tool to help affinity groups stay one move ahead of police during the NYC Republican National Convention protests of 2004).[82] The key point is that attribution and attention are important benefits of design processes, and they should be more equitably distributed. Innovation in media technologies, like all sociotechnical innovation, is an interplay between users and tool developers, not a top-down process. Social movements in particular have always been a hotbed of innovation in media tools and practices, in part because of the relationship between the media industries and social movement (mis)representation. Social movements, especially those led by marginalized communities, are systematically ignored, misrepresented, and attacked in the mass media, so movements often form strong community media practices, create active counterpublics, and develop media innovations out of necessity.[83] Social movement media innovations are later adopted by the broader cultural industries. Examples include TXTMob and Twitter, DIY livestreams from DeepDish TV to Occupy, and message encryption from Signal to WhatsApp. These stories have to be more widely told so that movements' contributions to the history of technology aren't erased. The last section of the chapter explores the importance of design scoping and framing, and critically analyzes how design challenges act as antipolitics machines. How do institutions frame and scope "problems" for designers to "solve" in ways that systematically render structural inequality, history, and community resistance invisible? Ultimately, the chapter maintains, we need a shift from deficit to asset-based approaches to design scoping; we also need community leadership in design processes during scoping and "challenge" definition phases of a design cycle, not only during the "gathering ideas" or "testing our solutions" phases.

Chapter 4 considers the question, "Where do we do design work?" Of course, design takes place everywhere, including in subaltern design sites and in marginalized communities. However, particular sites are valorized as ideal-type locations for design practices. The first part of this chapter explores the growing literature about design sites like hacklabs, makerspaces, fablabs, and hackathons—places where people gather to share skills, learn, design, prototype, make, and build using new technologies. Some scholars argue that originally hacklabs were explicitly politicized spaces at the intersection of social movement networks and geek communities.[84] Over time, startup culture and neoliberal discourses of individual mastery and entrepreneurial citizenship largely coopted hacklabs,[85] even as city administrators leveraged technofetishism to create municipal "innovation labs." This section also provides a critical analysis of the fablab network.

Next, the chapter interrogates the ideals, discourse, and practices of hackathons: What do people think hackathons do, and what really happens at hackathons? In what ways do they challenge and/or reproduce the matrix of domination? How might we imagine them as more intentionally liberatory and inclusive sites structured by design justice principles and practices? There has been a recent move toward intentional diversification of hacklabs, makerspaces, and hackathons, specifically along lines of gender, race, and sexual orientation. Examples include DiscoTechs (pioneered by the Detroit Digital Justice Coalition), CryptoParties, Trans*H4CK, #A11yCAN Hackathons, and the Make the Breast Pump Not Suck Hackathon and Policy Summit, among many others. In addition to the diversification of hacklab participants, the chapter concludes that design justice requires a broader cultural shift, back toward intentional linkage of these sites to social movement networks.

Chapter 5 is an extended reflection on critical pedagogies of design justice. It asks, "How do we teach and learn design justice?" It begins with a summary of the ideas behind critical pedagogy and popular education, based on work by Paulo Freire, bell hooks, and others. The chapter places these ideas in dialogue with constructionist design education theorists such as Seymour Papert and Mitchel Resnick, as well as the community technology pedagogy of Diana Nucera and the Detroit Community Technology Project, Maya Wagoner's Critical Community Technology Pedagogy, and Catherine D'Ignazio and Laura Klein's feminist pedagogy of data science, among others. In the second half of the chapter, I draw from my own experience teaching the Civic Media: Codesign Studio course at the Massachusetts Institute of Technology over the last six years. I synthesize lessons from the Codesign Studio case studies and consider them within the framework of the ten Design Justice Network Principles. The chapter ends with a reflection on the famous debates between W. E. B. DuBois and Booker T. Washington about the nature of education, and asks us to consider: What would it mean for institutional structures to support a community-based pedagogy of technology design? What are the challenges in an age of the neoliberalization of the educational system?[86] Is the aim of computing education to make all people good coders, or to make all coders good people?

The book ends with more questions than conclusions. "Directions for Future Work" describes the growing #TechWontBuildIt movement and asks, "What are some important directions for future design justice work?" It considers tensions between design justice processes and their outcomes, the role of Black feminist thought in design theory writ large, the paradox of pragmatic design, and the need for more specific design justice work in design domains like architecture, urban planning, industrial design, fashion design, and more. Next, it examines possible future areas for expanding the design justice framework, such as in project evaluation and impact assessment; guidelines, standards, codes, and laws; and the dynamics of unintended consequences. The chapter concludes with reflections on design justice and platform cooperativism, the need for more systematic resourcing for design justice sites, and possible institutional mechanisms to support design justice pedagogies. Finally, it points readers toward additional areas for research, and offers an invitation to join the growing community of design justice practitioners.

Limitations

Before we dive in, a brief note on the limitations of my approach: first, I believe that design justice is a framework that is applicable to all forms of design. However, my own practice and knowledge are limited to certain subfields, and I have drawn examples primarily from these. I encourage other scholars and practitioners to extend the design justice framework to other areas. In particular, I hope that others will explore the implications of design justice for industrial design, fashion design, and architecture, among other areas. I do not know these fields in depth and am not able to do them justice.

Another caveat: this is not a how-to manual. The Allied Media Projects Network Principles include the following: "Wherever there is a problem, there are already people acting on the problem in some fashion. Understanding those actions is the starting point for developing effective strategies to resolve the problem, so we focus on the solutions, not the problems. We emphasize our own power and legitimacy. We presume our power, not our powerlessness. We spend more time building than attacking."[87] Throughout this book, I have accordingly attempted to find a balance between critique of the ways design processes reproduce the matrix of domination and discussion of already existing design justice work. However, this is not a manual for practitioners. The Design Justice Network is producing excellent practical guides—for example, in its zine series, in the annual design justice track at the Allied Media Conference, via the network's website, and in other ways. I hope that at some point soon the network will produce a design justice methods kit; for now, I urge readers who are more interested in immediately putting design justice into practice in their own work to explore http://designjusticenetwork.org.

Overall, design justice, both as a conceptual framework and as a community of practice, provides a normative and pragmatic proposal for a liberatory approach to design. *Normative* because design justice practitioners feel that we have an ethical imperative to systematically advance democratic participation in, and community control of, all stages of design. We therefore work to center historically marginalized communities in design processes. *Pragmatic* because, at the same time, we believe that design that follows these principles can produce images, objects, products, and systems that work better for all of us.

There is already a growing design justice community: people and organizations who work to realize design justice principles in our daily practices. In the spirit of accountability to community-led processes, the Design Justice Network Principles appeared near the beginning of this introduction. The Design Justice Network describes these principles as a "living document" and plans to continue to develop them with practitioners. I urge you, gentle reader, to reflect on them, incorporate them into your own work, and continue to develop them. Let's build the theory, practice, and pedagogy of design justice together!

1 Design Values: Hard-Coding Liberation?

Figure 1.1
We All Belong Here. Poster art by Micah Bazant, 2017.

Technology is always a form of *social* knowledge, practices and products. It is the result of conflicts and compromises, the outcomes of which depend primarily on the distribution of power and resources between different groups in society.
—Judy Wajcman, *Feminism Confronts Technology*

Design is the process by which the politics of one world become the constraints on another.
—Fred Turner[1]

If your beta social network doesn't allow blocking abusers from jump, your beta social network was probably developed by white dudes. #ello
—@AngryBlackLady[2]

"Black, Muslim, Immigrant, Queer! Everyone is welcome here!"

"We're Queer! We're Trans! No Walls! No Bans!"

"Let's get free! We've all gotta pee!"

The chants bounce off of the brutalist concrete walls of Boston City Hall. It's Sunday, February 5, 2017, sixteen days since Donald Trump was sworn into office as president of the United States of America. A crowd of over one thousand queer and trans* folks, immigrant rights activists, and our family, friends, and allies has assembled at Government Center. Huge rainbow LGBTQI+ flags and smaller pink, blue, and white trans* flags flap in the frigid February air, alongside the black, red, and white banners of the antifascist action contingent. I climb the steps, carrying my *djembe*; I've been asked to help drum and lead chants during the march over to Copley Square. First, though, five of the protest organizers, each from a diverse intersection of queer, trans*, and other identities—Black, Latinx, Asian, immigrant, Disabled working-class—pass a microphone around and read a collective statement about the reasons for the mobilization:

Trump claims that he wants to protect the LGBTQ community from oppression (!!!) But … the unconstitutional #MuslimBan impacts millions of people including Queer and Trans people; border walls and militarization means more deaths in the desert, including QT deaths; expanding raids and detention and deportations affects all of us and especially increases the violence that Undocuqueer folks experience; the return of the Global Gag Rule undermines reproductive justice; attacks on Native sovereignty through reopening #KeystoneXL and #NoDAPL are attacks on all Native people including Two-Spirit people; threats to implement voter ID laws in all 50 states, because Trump lost the popular vote, mean disenfranchisement for marginalized voters; Trump's promise to take Stop and Frisk nationwide will target Black and Brown people in every community, and we know that Queer, Trans, and Gender Non-Conforming Black and Brown people are among those MOST targeted by racist police violence. … We're here for Black, Muslim, Native, Immigrant, Queer, Disabled, Women, POC communities, and for all those who live at the intersections of many of these communities at the same time! We are here for each other, because our liberation is bound up together. We see each other and we have got each other's backs. None of us are free until all of us are free! If we don't ALL get it: #SHUTITDOWN!"[3]

The Trans* and Queer Liberation and Immigrant Solidarity Protest was part of the massive wave of street mobilizations that took place in the winter and spring of 2017 in response to the election and the first actions of the Trump administration. It was organized in less than a week through the efforts of #QTPower, an ad hoc collective of Boston-based activists, lawyers, cultural workers, and community organizers that I participated in. The primary tools that we used to organize actions so quickly were face-to-face meetings and conference calls to plan key aspects of the event; Google docs to draft framing, language, demands, and logistical details; email, phone calls, and instant messaging to gather organizational cosponsors; and Facebook to promote the action. We also employed the symbolic power of solidarity imagery by Micah Bazant, a visual artist whose work has circulated widely through social movement networks over the last decade (we used figure 1.1, above, to promote our event).

By now, activist use of Facebook as a tool to help organize political protests is a story that has been widely told in both scholarly and popular writing. The best of these accounts complicate any simplistic narrative about the relationship between social media platforms and political protest activity. Zeynep Tufekci, a media scholar and public intellectual who studies the social impacts of technology, argues that tools like Facebook enable social movements to mobilize large numbers of people quickly around a simple broad demand, even when the movement lacks capacity to do much else.[4] Paolo Gerbaudo, a social movement scholar who studied the so-called Arab Spring, the Occupy movement, and the Spanish 15-M movement, describes how charismatic activists with large numbers of followers use Facebook and

Twitter to lead social movements through what he calls a *choreography of assembly*, without developing mechanisms of representative democratic decision making.[5] Ramesh Srinivasan, a scholar and design theorist who works with migrant and indigenous communities, cautions against the oversimplified US mass media narrative that tries to claim credit for the Arab Spring as an inevitable outcome of the introduction of social media platforms. Instead, he highlights the politics, history, organizations, and critical consciousness of local activists who made the uprisings and revolutions happen.[6] Communication scholars Moya Z. Bailey, Sarah Jackson, and Brooke Foucalt-Welles analyze how social movement networks leverage the affordances of Twitter in hashtag activism campaigns like #SayHerName, #GirlsLikeUs, and #MeToo.[7]

Other scholars of media, ICTs, and social movements, such as Emiliano Treré, Bart Cammaerts, and Alessandra Renzi, provide detailed discussion of the specific ways that the designed affordances of Facebook (and other social media platforms) enable and constrain activist use.[8] For example, as we organized the #QTPower actions, Facebook Events provided excellent tools for quickly circulating our call to action to thousands of people and for gauging interest through the built-in RSVP feature. During this period of heightened mobilization, event RSVPs ("I'm going," in Facebook terms) mapped more closely to real-world turnout than usual. Protest organizers shared with one another that events with thousands of RSVPs were likely to actually have thousands of people show up, compared with far smaller turnout ratios at other times.[9]

Yet Facebook in general, and Facebook Events specifically, provides terrible tools for the most important task of community organizers: to move people up the ladder of engagement.[10] After the #QTPower event was over, it was possible to share some additional information, photos, and feedback via the event page, but even as the event organizers we had no way to broadcast a message about our next move to all attendees. Posts to the event discussion only appeared for some people, subject to the opaque decision making of Facebook's News Feed algorithm. Of course, we had the option to pay Facebook a fee to make it more likely that protest attendees would find additional content from the mobilization in their feeds. Yet the platform design denied us the ability to do what we most wanted to do: in this case, contact all of the protest attendees (and those who had expressed interest in the event) and invite them to the next mobilization, scheduled for a few weeks later when the Trump administration announced a rollback of Title IX protections for trans* and gender-non-conforming students across the country.[11]

The poor fit between Facebook's affordances and basic activist needs partly explains the existence of an entire ecosystem of dedicated activist Constituent Relationship Management systems (CRMs), such as SalsaCommons, NationBuilder, and Action Network. These platforms, designed around the needs of community organizers and political campaigners, have built-in features, interface elements, and capabilities that match the core processes of building campaigns. They provide tools such as mass email lists, petitions, events, surveys, and fundraising, as well as ladder-of-engagement services such as activist performance tracking, list segmentation, and automated reminders and instructions. These types of tools are built into activist CRM platforms in part because the founders of these platforms typically have experience running campaigns and understand these needs, and also because most employ User-Centered Design (UCD) methods and agile development to continually improve the fit between platform affordances and user needs.

Yet such platforms remain niche services, used by only a relatively tiny group of professionalized campaigners. They typically cost money to use, often based on the number of contacts in the campaign database, and they require a significant investment of time and energy to learn. They will in all likelihood never be widely adopted by the vast majority of people who participate in social movements. Instead, most people, including social movement activists, organizers, and participants, use the most popular corporate social network sites and hosted services as tools to advance our goals. We work within the affordances of these sites and work around their limitations. We do this even when these tools are a poor fit for the specific task at hand, and even when their use exposes movement participants to a range of real harms.

Why do the most popular social media platforms provide such limited affordances for the important work of community organizing and movement building? Why is the time, energy, and brilliance of so many designers, software developers, product managers, and others who work on platforms focused on optimizing our digital world to capture and monetize our attention, over other potential goals (e.g., maximizing civic engagement, making environmentally sustainable choices, building empathy, or achieving any one of near-infinite alternate desirable outcomes)? Put another way, why do we continue to design technologies that reproduce existing systems of power inequality when it is so clear to so many that we urgently need to dismantle those systems? What will it take for us to transform the ways that we design technologies (sociotechnical systems) of all kinds, including digital interfaces, applications, platforms, algorithms, hardware, and infrastructure, to help us advance toward liberation?

Everyday Things for Whom? The Distribution of Affordances and Disaffordances under the Matrix of Domination

Let's begin with one of the core concepts of design theory: affordances. According to the Interaction Design Foundation, affordances are "an object's properties that show the possible actions users can take with it, thereby suggesting how they may interact with that object. For instance, a button can look as if it needs to be turned or pushed."[12] The term *affordances* was initially developed in the late 1970s by cognitive psychologist James Gibson, who states that "the affordances of the environment are what it offers the animal, what it provides or furnishes, either for good or ill."[13] It came to be influential in various fields following design professor William W. Gaver's much-cited article "Technology Affordances,"[14] and then it moved into even wider use in human-computer interaction following the publication of cognitive scientist and interface designer Donald Norman's *The Design of Everyday Things*.[15] For Norman, *affordance* refers to "the perceived and actual properties of the thing, primarily those fundamental properties that determine just how the thing could possibly be used."[16] For example, a chair *affords* sitting, a doorknob *affords* turning, a mouse *affords* moving the cursor on the screen and clicking at a particular location, and a touchscreen *affords* tapping and swiping.

The Design of Everyday Things is a canonical design text. It's full of useful insights and compelling examples. However, it almost entirely ignores race, class, gender, disability, and other axes of inequality. Norman very briefly states that capitalism has shaped the design of objects,[17] but says it in passing and never relates it to the key concepts of the book. Race and racism appear nowhere. He uses the term *women* only once, in a passage that describes the Amphitheatre Louis Laird in the Paris Sorbonne, where "the mural on the ceiling shows lots of naked women floating about a man who is valiantly trying to read a book."[18] Gay, lesbian, transgender: none of these terms appear. Disability is barely discussed, in a brief section titled "Designing for Special People." In this three-page passage, Norman describes the problems designers face in designing for left-handed people and urges the reader to "consider the special problems of the aged and infirm, the handicapped, the blind or near-blind, the deaf or hard of hearing, the very short or very tall, or the foreign."[19] He thus firmly subscribes to the individual/medical model of disability that locates disability in "defective" bodies and as a "problem" to be solved, rather than the social/relational model (that recognizes how society actively disables people with physical or psychological differences, functional limitations, or impairments through unnecessary exclusion, rather than taking action to meet their access needs[20]), let alone the disability justice model, created by Disabled B/I/PoC as they fight to dismantle able-bodied supremacy as a key axis of power within the matrix of domination.[21] Norman provides a single footnote about a multilingual voice message system, and another about typewriter keyboards and the English language.[22] In other words, the book is a compendium of designed objects that are difficult to use that provides key principles for better design, but it almost entirely ignores questions of how race, class, gender, disability, and other aspects of the matrix of domination shape and constrain access to affordances. Design justice is an approach that asks us to focus sustained attention on these questions, beginning with "how does the matrix of domination shape affordance perceptibility and availability?"

Affordance Perceptibility and Availability

First, we might ask whether any given affordance is *equally perceptible* to all people, or whether it systematically privileges some kinds of people over others. Gaver does recognize, but greatly downplays, the role that *standpoint* (in his terms, *culture*, *experience*, and *learning*) plays in determining affordance perceptibility. He acknowledges that culture and experience serve to "highlight" some affordances for a given user, but states that this is not "integral to the notion" of affordances.[23] However, there are much stronger claims to be made about the ways that standpoint shapes affordance perceptibility. For example, Gaver describes the carefully designed perceptual cues that reveal the affordances of scroll bars to a (abstracted, universalized) user. However, a person who is blind, visually impaired, or is interacting with a computer for the first time in their life will receive *few to none* of the benefits of these cues. Nor will the perceptual cue of a floppy disk icon located beside the Save option in a dropdown menu help someone who has never used a floppy disk understand the affordance on offer, at least until they have learned what it means. A person unfamiliar with the Roman alphabet will not benefit from the perceptual information offered by the text "Save," as anyone who has ever tried to use a computer with menus set to an unfamiliar language (let alone an unfamiliar alphabet!) will know. Affordance perceptibility also often differs for people who are colorblind, blind, or have visual impairments, or for people who are deaf or hard of hearing. Standpoint thus determines whether an affordance is *perceptible at all* to a given user, and affordance perceptibility is always shaped by standpoint (location within the matrix of domination); every affordance is more perceptible to some kinds of users than to others.

Second, in addition to perceptibility, design justice impels us to consider whether a given affordance is *equally available* to all people. For example, stairs (another example provided by Gaver) afford moving between two levels of a home for most people but deny this affordance to those whose type of mobility makes stairs difficult or impossible to use. For these users, stairs may provide a *perceptible but unavailable* affordance. An audible alert announcing the arrival of an instant message may enhance perception of the affordances of an instant message client for some users (those who are able to hear the alert, those who have the application minimized in the background, or those who are away from the computer while engaged in another task that occupies their visual attention), but offers no perceptual advantages to other users (those who are deaf or hard of hearing, who have their computers muted, who are in a very noisy workplace, etc.). An object's affordances are never equally *perceptible* to all, and never equally *available* to all; a given affordance is always more perceptible, more available, or both, to some kinds of people. Design justice brings this insight to the fore and calls for designers' ongoing attention to the ways these differences are shaped by the matrix of domination.

Disaffordances and Dysaffordances

As we have discussed, design affordances match perceptual cues with actions that can be performed with an object. In contrast, design *disaffordances* match perceptual cues with actions that will be blocked or constrained. In a paper about discriminatory design, philosopher of technology D. E. Wittkower provides many examples of disaffordances: a fence *disaffords* entry to a plot of land; a lock on a door *disaffords* entry without a key; and a fingerprint scanner on a mobile phone affords access to the phone's content for the owner, while it disaffords access to all others.[24] Wittkower also identifies *dysaffordances* (a subcategory of disaffordances), a term he uses for an object that requires some users to misidentify themselves to access its functions. For example, as a nonbinary person, I experience a dysaffordance any time I interact with a system, such as air-travel ticketing, that forces me to select either Male or Female to proceed. While a graduate student, Joy Buolamwini experienced the dysaffordances of facial detection technology, which failed to detect her dark-skinned face until she donned a white mask. This led her to systematically study bias in facial analysis technology and to found the Algorithmic Justice League.[25] Design justice asks us to constantly consider the distribution of affordances, disaffordances, and dysaffordances among different kinds of people.

For example, for Gaver, a doorknob *affords* turning, and "the interaction of a handle with the human motor system determines its affordances. When grasping a vertical bar, the hand and arm are in a configuration from which it is easy to pull; when contacting a flat plate pushing is easier."[26] Design justice, grounded in critiques developed by the disability justice movement, asks us to question the universalizing assumption that there is only one configuration of the human motor system. Instead, there are many configurations; some will be privileged (supported) by a vertical bar as a mechanism to pull a door, and others will find that particular combination of object and action difficult or nearly impossible: an affordance for some is a disaffordance for others. For example, a small child might find it extremely difficult to open a door based on pulling a vertical bar at adult chest height; a more appropriate design solution for them (if the goal is to enable door-opening) might be a door that swings in both directions. This design is also common in scenarios in which users are not expected to have the use of their hands and arms for door-opening—for example, in doors to restaurant kitchens, where waitstaff's hands and arms are often occupied with plates and dishes. The point of a design justice analysis here is not to impose a single, "best" design solution, but to recognize that affordances, disaffordances, and dysaffordances privilege some people over others.[27]

Both the perception and availability of any given affordance, as well as disaffordances and dysaffordances, are shaped, in part, by the matrix of domination.

Intention and Impact

Most designers today do not intend to systematically exclude marginalized groups of people. However, power inequalities as instantiated in the affordances and disaffordances of sociotechnical systems may be intentional or unintentional, and the consequences may be relatively small, or they may be quite significant. For example, technology writer Sara Wachter-Boettcher describes how default characters in "endless runner" game apps appear as "male" 80 percent of the time and "female" just 15 percent of the time (the other 5 percent are nonhuman characters). Default avatars in this game genre are an affordance that systematically privileges masc identifying people over femme identifying people, although this difference is likely unintentional and the impact is relatively small.[28] In the built environment, perhaps the most famous (and controversial) example of discriminatory design is scholar of science, technology, and society Langdon Winner's story about NYC urban planner Robert Moses's overhead passes that (may have) blocked public buses from reaching the Rockaway beaches.[29] For Winner, this illustrates how planners can structure racism into the built environment. Some have questioned whether this designed constraint was intentionally racist, while others have questioned whether bus traffic was ever, in fact, constrained at all.[30] Winner's broader point, however, developed throughout his many articles and several books on the topic, is that technologies embody social relations (power). Through a design justice lens, we might say more specifically that under neoliberal multicultural capitalism, most of the time designers unintentionally reproduce the matrix of domination (white supremacist heteropatriarchy, capitalism, and settler colonialism).

Most designers, most of the time, do not think of themselves as sexist, racist, homophobic, xenophobic, Islamophobic, ableist, or settler-colonialist. Some may consider themselves to be capitalist, but few identify as part of the ruling class. Many feel themselves to be in tension with capitalism, and many even identify as socialist. However, design justice is not about intentionality; it is about process and outcomes. Design justice asks whether the affordances of a designed object or system disproportionally reduce opportunities for already oppressed groups of people while enhancing the life opportunities of dominant groups, independently of whether designers intend this outcome.

Of course, sometimes designers do intentionally design objects, spaces, and systems that are explicitly oppressive. For example, as surveillance scholar Simone Browne excavates in her brilliant text *Dark Matters: On the Surveillance of Blackness*, the designers of slave ships used for the transatlantic slave trade intentionally designed, planned, and participated in the construction of ships, ship modifications, and cargo hold instruction manuals to afford the transport of the greatest possible number of enslaved African human beings per voyage.[31] Designers of prisons and detention centers today also participate in explicitly oppressive projects. One architectural firm, HOK Justice Group, boasts that it has designed prison and detention facilities with more than one hundred thousand beds.[32] Most recently, there is a debate within the American Institute of Architects over whether architects should participate in the redesign of immigration detention facilities to improve conditions for people who are held in them, or whether architects have an ethical responsibility to boycott such work entirely.[33] Nearly two hundred design, engineering, and construction firms bid for contracts to build the Trump administration's xenophobic border wall.[34] In the final chapter of this book, we will return to this conversation in the form of the #TechWontBuildIt movement.

Discriminatory Design

Historian and scholar of race, gender, and science and technology studies Ruha Benjamin defines *discriminatory design* as the normalization of racial hierarchies within the underlying design of sociotechnical systems.[35] Benjamin uses the example of the *spirometer*, a device meant to assess lung capacity: because early pulmonologists believed that "race" determined lung capacity, spirometers were built with a "race correction" button to adjust measurements relative to an expected norm. In a 1999 class action lawsuit from fifteen thousand asbestos workers against their employer, this made it difficult for Black workers to qualify for workers compensation because "they would have to demonstrate worse lung function and more severe clinical symptoms than those for white workers due to this feature of the spirometer, whose developer, Dr. John Hutchinson, was employed by insurance companies in the mid-1800s to minimize payouts."[36] For Benjamin, the reproduction of ideas about race and racial difference in the hardware, software, and operation of the spirometer is an example of how science and technology are central sites in modern "racecraft." In Benjamin's (2019) book *Race After Technology*, she expands her arguments about discriminatory design, analyzes multiple examples, and links the current conversation about machine bias to an analysis of systemic racism. Benjamin demonstrates the ways that racial discrimination becomes hidden, buried, and "upgraded" through the deployment of new technologies that hide oppression in the default settings and that mask racist logic as consumer choice or desire.[37] In the recent edited volume *Captivating Technology: Race, Carceral Technoscience, and Liberatory Imagination in Everyday Life* (2019), Benjamin gathers authors who explore how sociotechnical systems developed by the carceral state to support racial hierarchy and social control, such as electronic ankle monitors and predictive policing algorithms, have now been deployed in ever-more domains of life, such as schools, hospitals, workplaces, and shopping malls.[38]

To take another example, an article by Soraya Chemaly for *Quartz* focuses on new technology that is designed for men, with the assumption that the user will be male.[39] One study described by Chemaly found that virtual assistants Siri, Cortana, Google Assistant, and S Voice were all able to respond to queries about what to do in case of heart attack or thoughts of suicide, but none recognized the phrases "I've been raped" or "I've been sexually assaulted,"[40] despite the high rates of rape, sexual assault, and intimate partner violence experienced by women and femmes.[41] As Chemaly notes:

> The underlying design assumption behind many of these errors is that girls and women are not "normal" human beings in their own right. Rather, they are perceived as defective, sick, more needy, or "wrong sized," versions of men and boys. When it comes to health care, male-centeredness isn't just annoying—it results in very real needs being ignored, erased or being classified as "extra" or unnecessary. To give another, more tangible example, one advanced artificial heart was designed to fit 86% of men's chest cavities, but only 20% of women's ... the device's French manufacturer Carmat explained that the company had no plans to develop a more female-friendly model as it "would entail significant investment and resources over multiple years."[42]

Discriminatory design often operates through standardization. Everything from the average size and height of a seat in a car to the size of "ergonomic" finger depressions in tool handles to the size of frets on a guitar were all initially developed based on statistical norms that privilege one-third world male bodies (*one-third world* is feminist scholar Chandra Mohanty's reformulation of the dated and hierarchical term *first world*).[43] Such discriminatory standards are not exceptions; rather, they shape technologies in nearly every field, such as transportation, health, housing, clothing, and more.[44]

What's more, although design that discriminates based on race and/or gender is often seen as problematic, social norms under capitalism do support systems design that intentionally reproduces class-based discrimination. For example, the intended purpose of a predictive algorithm used by the credit industry to determine home loan eligibility is to afford the loan officer a heightened ability to discriminate between those who are likely to be able to make loan payments and those who are likely to fall behind. Such a tool, by definition, promotes class-based discrimination, and when it does so, it is seen to be doing its job. However, when it discriminates based on a single-axis characteristic (race *or* gender *or* disability) that is explicitly protected by the law, then it is said to be biased.

In general, predictive algorithms often support and afford racist decision making. This happens constantly, although today's algorithm developers (unlike the designers of redlining policies in the past) do not usually use race intentionally as a variable to lower loan-eligibility scores. Instead, algorithm developers and the banks that employ them use machine-learning techniques to produce risk constructs that don't have clearly identified real-world variables but are actually stand-ins for race, class, gender, disability, and other axes of oppression. The erasure of history and the failure to consider intersectional structural inequality underpins the pretense of "fairness" in such decision-making support systems, even as they work to reproduce the matrix of domination.

Disaffordances as Microaggressions

Discriminatory design, or the unequal distribution of affordances and disaffordances, may also be experienced as microaggressions by individuals from marginalized groups. Racial microaggressions are "brief and commonplace daily verbal, behavioral, or environmental indignities, whether intentional or unintentional, that communicate hostile, derogatory, or negative racial slights and insults toward people of color."[45] Recently, research has extended the study of microaggressions to online interactions, such as in chat rooms, on social media platforms,[46] and in real-time online multiplayer games.[47] Microaggressions reproduce the matrix of domination; reaffirm power inequalities; generate a climate of tension within organizations and communities; produce physical, cognitive, and emotional shifts in targeted individuals; and, over time, reduce both quality of life and life expectancy for people from marginalized groups.[48]

Microaggressions are (often unintentional) expressions of power and status by individuals from dominant groups, performed against individuals from marginalized groups who may also frequently experience far more severe manifestations of oppression, such as physical violence, attack, rape, or murder, as well as severe forms of institutional inequality such as discriminatory exclusion from access to employment and housing. In many contexts, an individual experiencing a microaggression has no way of knowing whether it is about to escalate into something more severe. For example, as a trans* femme individual, I was walking home after dinner one evening last year. A car with tinted windows slowed down, cruised alongside me, and a deep voice from inside yelled out, "What is it? Is it a girl or a boy? Would you fuck that?" I had no way of knowing at that moment whether the aggression would remain at the level of verbal abuse, or if the situation was going to escalate to physical violence.[49] Thus, although microaggressions are often read as relatively harmless and usually unintentional expressions of racial and/or gender bias, we can also understand them as small-scale, pervasive, daily, and constant performance of power. Metaphorically, they are the fabric, molecules, or smallest-level building blocks that constantly reproduce, replenish, and strengthen larger systems of oppression. They also serve to constantly put marginalized groups "in their place."

Looking at biased systems through the lens of microaggressions means trying to understand the impact on individuals from marginalized groups as they encounter, experience, and navigate these systems daily. For example, a Black person might experience a microaggression if their hands do not trigger a hand soap dispenser that has been (almost certainly unintentionally) calibrated to work only, or better, with lighter skin tones. This minor interruption of daily life is nevertheless an instantiation of racial bias in the specific affordances of a designed object: the dispenser affords hands-free soap delivery, but only if your hands have white skin. The user is, for a brief moment, reminded of their subordinate position within the matrix of domination.[50]

For many people from marginalized groups, the ways that the matrix of domination is both reproduced by and produces designed objects and systems at every level—from city planning and the built environment to everyday consumer technologies to the affordances of popular social media platforms—generates a constant feeling of alterity. The sentiment that "this world was not built for us" is regularly expressed in intellectual, artistic, poetic, musical, and other creative production by marginalized groups. It is a common refrain, for example, in Afrofuturist work. Consider Jamila Woods's lyrics: "Just cuz I'm born here, don't mean I'm from here; I'm ready to run, I'm rocket to sun, I'm waaaaay up!"[51] Experiences of design microaggression are proximally based on a particular interaction with an object or system. However, they instantiate, recall, and point to much larger systems, histories, and structures of oppression within the matrix of domination. Even if only for a moment, the user is "put in their place" through the interaction.

Attention to the real, cumulative, and lasting effects of what seem (to those who do not experience them) like minor microaggressions should not displace attention to the many ways that biased affordances often have quite significant and life-altering effects on marginalized groups of people, as in biased pretrial detention, sentencing, or home loan algorithms. Instead, we might say that design constantly instantiates power inequality via technological affordances, across domains, in ways both big and small. Seemingly minor instances may be experienced by individuals from marginalized groups as microaggressions, and these can have significant impacts as they accumulate over a lifetime. A design justice framework can help shift the conversation so that each time an instance of racial or gender bias in technology design causes a minor scandal, it will not be seen as an "isolated incident," a "quirky and unintentional mistake," or even used as fodder for an argument that "someone on the design team must have been racist/sexist." Instead, design justice argues that such moments should be read as the most visible instances of a generalized and pervasive process by which existing technology design processes systematically reproduce (and are reproduced by) the matrix of domination.

Related Approaches: Value-Sensitive Design, Universal Design, Inclusive Design

Design justice builds on, but also differs in important ways from, related approaches such as value-sensitive design, universal design, and inclusive design. The second part of this chapter briefly explores these related frameworks, both in terms of shared concepts and in terms of differences in theory and practice.

Value-Sensitive Design

Science and technology scholars have long argued that tools are never neutral and that power is reproduced in designed objects, processes, and systems.[52] In the 1990s, in an effort to address unintentionally biased design in computing systems, information scientists and philosophers Batya Friedman and Helen Nissenbaum developed the concept of *value-sensitive design* (VSD).[53] In the earliest and most widely cited book on this approach, *Human Values and the Design of Computer Technology*, Friedman and Nissenbaum examine bias in computer systems and propose methods for the practice of VSD.[54] They analyze seventeen computer systems from varied fields, expose instances of bias, and categorize them into three groups: preexisting bias, technical bias, and emergent bias. In *preexisting bias*, bias that exists in broader society, culture, and/or institutions is reproduced in the computer system, either intentionally or unintentionally, by systems developers. For example, graphical user interfaces typically embody a preexisting bias against vision-impaired people because the designers do not consider their existence at all, not because they consciously decide to exclude them.[55] In *technical bias*, some underlying aspect of the technology reproduces bias; for example, the poor performance of optical sensors on darker-skinned people. In *emergent bias*, a system that may not have been biased given its original context of use or original user base comes to exhibit bias when the context shifts or when new users arrive—for example, Tay, the Microsoft chatbot that was trained to be sexist and racist by Twitter users.[56] VSD does not believe that most designers are intentionally racist, sexist, or malicious. Instead, this approach emphasizes that many mechanisms that introduce unintentional bias are at play. These include "unmarked" end users, biased assumptions, universalist benchmarks, lack of bias testing, limited feedback loops, and, most recently, the use of systematically biased data sets to train algorithms using machine-learning techniques.[57]

Designers often assume that "unmarked" users occupy the most privileged position in the matrix of domination (a point discussed further in chapter 2). Science and technology scholar Ruha Benjamin has written about how normative assumptions lead to what she calls the "New Jim Code—the employment of new technologies and social design that reflect and reproduce existing inequities but which we assume are more objective or progressive than discriminatory systems of a previous era."[58] My personal experience of design teams in many contexts is that designers often assume users to be white, male, abled, English-speaking, middle-class US citizens, unless specified otherwise. Unfortunately, this experience is supported by research. For example, Huff and Cooper (1987) found that designers of educational software for children assumed the user to be male, unless it was specified that the users were girls.[59] Other studies demonstrate that even designers from marginalized groups often make the same normative assumptions about unmarked users.[60] In the United States, designers tend to assume the user has broadband internet access, unless it is specified that they don't; that the user is straight, unless it's specified that the user is LGBTQ; that they are cisgender, unless it's specified that they are nonbinary and/or trans*; that they speak English as a first language, unless it's specified otherwise; that they are not Disabled, unless specified that they are; and so on.

Although much of designed bias is unintentional, Nissenbaum and Friedman also ask, "What is the responsibility of the designer when the client wants to build bias into a system?" They conclude that systems should be evaluated for "freedom from bias" and that such evaluation should be incorporated into standards, curriculum, and society-wide testing: "Because biased computer systems are instruments of injustice ... we believe that freedom from bias should be counted among the select set of criteria according to which the quality of systems in use in society should be judged. ... As with other criteria for good computer systems, such as reliability, accuracy, and efficiency, freedom from bias should be held out as an ideal."[61]

VSD provided an important shift in design theory and practice. However, design justice seeks more than "freedom from bias." For example, feminist and antiracist currents within science and technology studies have gone beyond a bias frame to unpack the ways that intersecting forms of oppression, including patriarchy, white supremacy, ableism, and capitalism, are constantly hard-coded into designed objects, platforms, and systems.[62] STS scholars and activists, such as those affiliated with the Center for Critical Race and Digital Studies,[63] have explored these dynamics across many design domains, from consumer electronics to agricultural technologies, from algorithm design in banking, housing, and policing to search engines and the affordances of popular social media platforms. To take one recent example, the organizing that took place around Facebook's "real name" policy illustrates how white supremacy and settler colonialism become instantiated in sociotechnical systems. As feminist blogger and cartoonist Alli Kirkham notes, "Native Americans, African Americans, and other people of color are banned disproportionately because, to Facebook, a 'real' name sometimes means 'traditionally European.'"[64] This happens, in part, because the algorithms that are used to flag likely "fake" names were trained on "real name" datasets that overrepresent European names, using machine-learning and natural-language-processing techniques. A Native American user thus may experience a microaggression if their name is flagged as fake by Facebook's Eurocentric fake name algorithm. This microaggression may "only" be a small inconvenience in the course of the person's day, yet it symbolically and materially invalidates the legitimacy of their identity. The system instantiates a new, tiny instance of the erasure of Native peoples (genocide) under settler colonialism. After significant pushback from various communities, Facebook claims that it has modified the algorithm to correct for this bias. However, no external systematic study has yet verified whether the situation has improved for those with non-European names.

Together with the fight against hard-coded Eurocentricity, there have been extensive efforts to push back against various aspects of Facebook's gender normativity.[65] The LGBTQ community, and drag queens in particular, successfully organized to force Facebook to modify its real name policy. Many LGBTQ folks choose to use names that are not our given names on social media platforms for various reasons, including a desire to control who has access to our self-presentation, sexual orientation and/or gender identity (SOGI). For many, undesired "outing" of a nonhetero- and/or cis-normative SOGI may have disastrous real-world consequences, ranging from teasing, bullying, and emotional and physical violence by peers to loss of family, housing instability, and denial of access to education, among others. For years, Facebook systematically flagged and suspended accounts of LGBTQ people who it suspected of not using real names, especially drag queens—and drag queens fought back. After several prominent drag queens began to leave the hegemonic social network for startup competitor Ello, Facebook implemented some modifications to its real name flagging and dispute process and instituted a new set of options for users to display gender pronouns and gender identity, as well as more fine-grained control over who is able to see these changes. However, as scholar of data, information, and ethics Anna Lauren Hoffman notes, the diverse gender options only apply to display; on the back end, Facebook still codes users as male or female.[66]

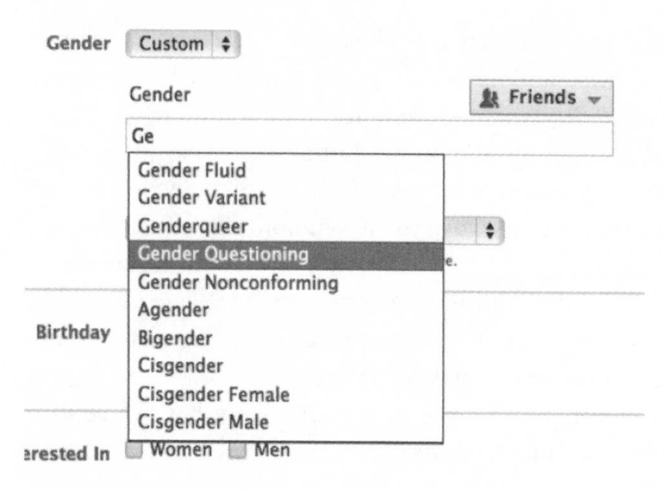

Figure 1.2
Screen capture of Facebook gender options. *Source:* Oremus 2014.

These are examples of how dominant values and norms are typically encoded in system affordances—in this case, assumptions about names, pronouns, and gender that were built into various aspects of Facebook's platform. They also demonstrate how, through user mobilization, platforms can, to some degree, be redesigned to encode alternative value systems. We need to develop many more case studies of user activism that targets values-laden elements of system design.

VSD advocates have also proposed tools to help designers incorporate the approach in practice. For example, Nissenbaum, Howe, and game designer Mary Flanagan suggest a library of value analyses to be used by designers to quickly develop functional requirements.[67] They also note that whether a particular design embodies the intended values is, in some cases, amenable to empirical inquiry. For example, they explore a hypothetical medical records system intended to promote user privacy through a multilevel permission system. In the thought experiment, the system fails to promote privacy in practice because users generally neglect to change default permissions, thereby widely exposing sensitive data—an outcome counter to the value intended by the designers. VSD proponents argue not only that technical artifacts embody values but also that it is possible for designers to deliberately design artifacts to embody a set of values that they choose. In a recently published book that provides an overview of three decades of VSD, Batya Freidman and David Hendry highlight key elements of the approach. They emphasize that VSD takes an interactional stance to technology and human values; that the values of various stakeholders implicated in the design should be considered by the designers; that values may be in tension with one another; that technology co-evolves with social structures; that we need to design for multi-lifespans; and that VSD emphasizes progress, not perfection.[68] VSD has been an important intervention in the design of computing systems. I will return below to some of the differences between design justice and VSD.

Disability and Universal(ist) Design?

In parallel to VSD, and with significant cross-pollination, over the last fifty years many designers have taken part in a long, slow shift toward deliberate design for accessibility. Historian Sarah Elizabeth Williamson describes how the disability rights movement worked for decades to transform discourse, policy, design, and practice, ultimately encoding rights to accessibility at multiple levels, including federal policy that governs architecture, public space, software interface design, and more.[69] Committed activists were able to accomplish many of these changes across multiple design fields, as documented by art and design historian Bess Williamson, among many others.[70] In computing, over time, a body of knowledge, examples, software libraries, automated tests, and best practices has grown along with a community of practice. Disabled people and their allies and accomplices implemented alternative interfaces such as text to speech; fought for engineering, architectural, and building standards to enable wheelchair access; convinced federal regulators to mandate closed captions in broadcast media; and much, much more.[71] Standardization steadily lowered implementation costs. At the same time, a legal regime was put in place that required designers in many fields to implement accessibility best practices, as Aimi Hamraie writes in their recent book *Building Access: Universal Design and the Politics of Disability*.[72]

This is not at all to imply that design practices now fully reflect a normalized concern for accessibility or incorporation of disability rights, let alone a disability justice analysis. For example, communication scholar Meryl Alpert has recently demonstrated that communication technologies meant to "give voice to the voiceless" continue to reproduce intersectional structural inequalities.[73] At the same time, disability justice underpins real gains. Alison Kafer's brilliant book *Feminist, Queer, Crip* draws from the history and practice of environmental justice, reproductive justice, disability justice, trans* liberation, and other movements to reimagine a radically inclusive world of Crip futures.[74] For example, trans* and GNC people, some abled and some who identify as *Crip* (in a move that reclaims the pejorative "cripple" as an in-group term of pride), are simultaneously challenging ableist spaces and the sociotechnical reproduction of the gender binary by struggling for (and in many cases winning) the implementation of gender-neutral, accessible bathrooms in schools, universities, public buildings, and private establishments across the country and around the world. Certainly, there is a possible future where gender-neutral, accessible bathrooms are standardized in most architectural plans, as well as mandated by law, at least in all public buildings and spaces. Along the same lines, scholars and activists like Heath Fogg Davis are pushing back against both public and private information systems and user interface design that regards gender as binary, that requires self-identification as Male or Female via dropdown menus, and that fails to recognize the gender identity or pronouns of system users.[75]

The history of design and disability activism provides the cornerstone for design justice. First, this history teaches us that it is indeed possible for a social movement to impact design policy, processes, practices, and outcomes in ways that are very broad, deep, and long-lasting. Disability rights and disability justice activists have changed federal policy, forced the adoption of new requirements in a wide range of design processes, altered the way many designers practice their craft, and significantly changed the quality of life for billions of people, not only for those who presently experience disabilities or identify as Disabled. Design justice is deeply intertwined with the disability justice movement and cannot exist apart from it (in chapter 2 we will return to additional discussion of disability justice).

In large part due to the efforts of Disabled activists, an approach known as *universal design* (UD) has gained reach and impact over the last three decades. UD emphasizes that the objects, places, and systems we design must be accessible to the widest possible set of potential users. In the 1990s, the Center for Universal Design at North Carolina State University defined UD as "the design of products and environments to be usable by all people, to the greatest extent possible, without the need for adaptation or specialized design."[76] For example, following UD principles, we need to add auditory information to crosswalk signals so that they will also be useful for Blind people and for anyone who has difficulty seeing or processing visual indicators. UD principles have led to real and significant changes in many design fields. However, as Aimi Hamraie has described, there is a tension between UD and disability justice approaches.[77] UD discourse emphasizes that we should try to design for everybody and that by including those who are often excluded from design considerations, we can make objects, places, and systems that ultimately function better for *all* people. Disability justice shares that goal, but also acknowledges both that some people are always advantaged and others disadvantaged by any given design, and that this distribution is influenced by intersecting structures of race, class, gender, and disability. Instead of masking this reality, design justice practitioners seek to make it explicit: we prioritize design work that shifts advantages to those who are currently systematically disadvantaged within the matrix of domination.

Inclusive Design

One group that has worked steadily to advance design practice that is not universalizing is the Inclusive Design Research Centre (IDRC). IDRC defines *inclusive design* as follows: "design that considers the full range of human diversity with respect to ability, language, culture, gender, age and other forms of human difference."[78] The IDRC's approach to design recognizes human diversity, respects the uniqueness of each individual, and acknowledges that a given individual might experience different interactions with the same design interface or object depending on the context. In addition, this group also sees disability as socially constructed and relational, rather than as a binary property (disabled or not) that adheres to an individual. Disability is "a mismatch between the needs of the individual and the design of the product, system or service. With this framing, disability can be experienced by anyone excluded by the design. ... Accessibility is therefore the ability of the design or system to match the requirements of the individual. It is not possible to determine whether something is accessible unless you know the user, the context and the goal."[79]

The group of designers and researchers who use this approach call for "one size fits one" solutions over "one size fits all." At the same time, they acknowledge that "segregated solutions" are technically and economically unsustainable. They argue that, at least in the digital domain, adaptive design that enables personalization and flexible configuration of shared core objects, tools, platforms, and systems provides a path out of the tension between the diverse needs of individual users and the economic advantages of a large-scale user base.[80]

Retooling for Design Justice

A paradigm shift to design that is meant to actively dismantle, rather than unintentionally reinforce, the matrix of domination requires that we retool. This means that there is a need to develop intersectional user stories, testing approaches, training data, benchmarks, standards, validation processes, and impact assessments, among many other tools. Yet the idea that we need to retool is sure to meet with great resistance. Physicist and philosopher of science Thomas Kuhn famously described how each scientific paradigm develops along with a widely deployed and highly specialized apparatus of experimentation, testing, and observation. These fixed costs reduce the likelihood of paradigm shift, absent a growing crisis where the current paradigm is unable to effectively account for discrepancies with the observed world. As Kuhn remarks: "As in manufacture, so in science—retooling is an extravagance to be reserved for the occasion that demands it."[81] As in manufacturing and in science, so in design: an intersectional critique of the ways that current design practices systematically reproduce the matrix of domination ultimately requires not only more diverse design teams, community accountability, and control, as we will explore in chapter 2, but also a retooling of the methods that shape so many design domains under the current universalist paradigm. That shift, however, will not come unless and until a large number of designers (and design institutions) become convinced that equitable design outcomes are a goal that is important enough to warrant retooling. It is my contention that this will only happen through organized, systematic efforts to demand design justice from a wide coalition of designers, developers, social movement organizations, policymakers, and everyday people. This section explores how a design justice analysis might help to rethink specific techniques and tools that designers use every day.

Make Me Think: Differential Cognitive Load

One of the most important goals in HCI, in particular for UI design, is to reduce the user's cognitive load to a minimum. Put simply, people should not have to think too hard to use computers to perform desired tasks. This imperative provides the title of designer Steve Krug's book *Don't Make Me Think*, sometimes known as "the Bible of interface design."[82] The book is a clearly written rundown of best practices in web usability. Unfortunately, in it, the imagined user is "unmarked" and universalized. Terms like *race*, *class*, and *gender* never appear. Somewhat surprisingly, the term *multilingual* is absent, and there is only one quick reference to a UI that requires language selection. Krug does devote a section to accessibility, but mostly to note that adhering to accessibility standards is often a legal requirement and that most sites can be made accessible after the fact without much effort.[83]

Taking a design justice approach, with attention to the distribution of benefits and burdens, we might ask: Is it always (or ever) possible to reduce cognitive load for *all* users simultaneously? Perhaps not. Instead, designers constantly make choices about *which users to privilege* and which will have to do more work. UI decisions distribute higher and lower cognitive loads among different kinds of people. The point is not that it's wrong to privilege some users over others; the point is that these decisions need to be made explicit.

Default language settings provide a simple example. In web applications design in the United States, if the default interface language is US English, there will be a higher bounce rate from, for example, monolingual Spanish speakers.[84] Providing an initial page that requires the user to make a language selection might reduce the bounce rate for this group and over time build a more multilingual user community on the site. However, this will also reduce overall traffic due to a loss of English-only users who don't want to click through the language-selection screen. On the other hand, if we choose to default to US English (as most sites in the United States do), we may lose site visitors who prefer Spanish (about forty-one million people in the United States speak Spanish at home).[85] What's more, because design decisions privilege one group of users over another, we shape the user base to conform to our (implicit or explicit) assumptions. Future A/B testing processes will be skewed by our existing user base, leading us to continually make decisions that reinforce our initial bias (A/B testing is discussed in more depth in the next section). For example, a test between Spanish-language menus and English-language menus will then be more likely to result in favorable results for the latter. In other words, initial design decisions about who to include and exclude produce self-reinforcing spirals.

Empirical studies support a strong critique of the idea that the same design is "best" for all users. For example, Reinecke and Bernstein found that most users preferred a user interface customized according to cultural differences. They note that it is not possible to design a single interface that appeals to all users; they argue instead for the design of "culturally adaptive systems."[86] Indeed, web designers increasingly talk about culturally adaptive and personalized systems and hope to shift toward providing personalized experiences for each user based on what they know about them. On the one hand, this approach has real potential to escape the reproduction of existing social categories as variables that are used to shape experience; it may destabilize existing social categories and replace them with truly personalized, behavior-driven user experience (UX) and UI customization. However, in practice this approach also leads to the reproduction and reification of existing social categories through algorithmic surveillance, tracking users across sites, gathering and selling their data, and the development of filter bubbles (only showing users content that we believe they are comfortable with).

Universalization erases difference and produces self-reinforcing spirals of exclusion, but personalized and culturally adaptive systems too often are deployed in ways that reinforce surveillance capitalism.[87] Design justice doesn't propose a "solution" to this paradox. Instead, it urges us to recognize that we constantly make intentional decisions about which users we choose to center and holds us accountable for those choices. Community accountability, control, and ownership of design processes is the topic of chapter 2.

A/B Tests and Denormalizing the "Universal User"

A/B testing is one of the most widely used methods for making design decisions. In A/B testing, users arrive at a web page or application screen and are randomly assigned to one of two (or more) versions of the same interface. Most elements are held constant, but one element (e.g., the size of a particular button) is varied. Designers then carefully observe and measure user interactions with the page, with an emphasis on key metrics such as time to complete a task. Whichever version of the interface performs better according to this key metric is then adopted. This approach has led to vast improvements in common web application UI and UX.

However, A/B testing is also nearly always deployed within a universalist design paradigm. For example, companies that operate platforms with billions of users, such as Google or Facebook, A/B test everything from the color of interface elements to major new features, from individual content items to recommendation algorithms. Based on the results, they roll out new changes to users. The underlying universalizing assumption is that A/B testing on existing users always results in a clear winner from the perspective of efficiency, reduced cognitive load, and user satisfaction, as well as (most importantly) profitability. The results of randomized A/B testing, it is assumed, will apply to all users. The change can be deployed, and the world will be a better place—or at least the firm will be a more profitable firm. However, what is A/B testing actually for? A/B testing is widely seen as leading inexorably to "better UX" and "better UI." But a question must be asked: Better for whom? Absent this question, A/B testing reproduces structural inequality through several mechanisms.

First, we should critique (trouble, queer, or denormalize) the assumption that A/B testing is always geared toward improving UX, for the simple reason that it is actually geared toward increasing the decision-making power of the product designer. The goals of the product designer are often in sync with the goals of many users, but this is not necessarily the case. For example, a product owner who wants to encourage users to share more personal information might A/B test various ways of encouraging (or requiring) users to do so. This is further complicated by the reality that product design decisions in medium to large firms are not necessarily made by the product designer. Instead, key decisions are frequently made further up the management chain. In this way, designers who may prefer a decision that would benefit users are often overruled by project managers or executives who prioritize profits. However, for the purposes of the present line of argument, this distinction is not important.

Second, we might destabilize the underlying assumption that what is best for the majority of users is best for *all* users. To take a simple example, consider a UI for personal profile creation on a university admissions portal. The site designer is required, for institutional diversity metrics, to request the race or ethnicity of the applicant. The designer is deciding how to implement the race/ethnicity selection process in as few clicks as possible to improve UX. In one version of the page (call it A), there is a default race/ethnicity set to White, Non-Hispanic. In a second version of the page (call it B), there is no default set, so the user must select their own race/ethnicity from a list. Now, keep in mind that the university applicant pool will reflect our broken and structurally unequal K–12 educational system, so the users of the site are disproportionally white. In a simple A/B test, the majority of (white) users would have a smoother experience, with fewer clicks required, under option A. However, can we therefore say that option A is the "best" option for user experience? In this case, what is best for the majority of current site visitors (set the default to White) produces an unequal experience, with the ever-so-slightly more time-consuming experience (additional clicks) reserved for PoC, who may also experience a microaggression in the process. Although our hypothetical "default to white" race/ethnicity dropdown is rarely implemented because of widespread sensitivity to such a blunt reminder of ongoing racial disparity, the same underlying principle is constantly used to develop and refine UX, UI, and other elements of sociotechnical systems.

How might we rethink A/B testing through a design justice lens? In some cases, it may not be a technique we can use. But in others, we may be able to compare responses from intersectional user subgroups. To generalize: imagine testing design options for an app with different kinds of users—for example, a group of Black women, a group of Black men, a group of white women, and a group of white men. If the design team sees statistically equivalent preferences from all groups, they may conclude that the design decision does not privilege one group over another. On the other hand, if the preferences of these different groups diverge, the design team must then discuss and intentionally decide what to do: if, say, both groups of women prefer one design, but both groups of men prefer another, the design team will have to make a decision about whose preferences to privilege.

Intersectional Benchmarks

Unfortunately, most design processes do not yet systematically interrogate how the unequal distribution of user experiences might be structured by the user's position within intersecting fields of race, class, gender, and disability. Design justice proposes the normalization of these types of questions and their adoption as a key aspect of all types of design work. At the moment, other than ADA compliance, questions of bias typically only surface when systems obviously fail some subset of raced and/or gendered users—for example, soap dispensers with higher error rates for darker skin, or cameras that don't recognize eyes without epicanthic folds as open.[88] Rather than understand these types of cases as marginalia, we might consider how they point to fundamental underlying problems of unexamined validation failure that are currently "baked-in" to most design processes.

A paradigm shift to a design justice approach replaces universalizing assumptions about test validity with an array of intersectional validation tests. This requires significant changes to existing instrumentation and product-testing processes. Consider the hand soap dispenser. Prior to the release of a commercial product, product engineers subject prototypes to a range of tests; these tests must typically be met at certain thresholds. In modern product design methods, they are likely to be couched in terms of *user stories* that must be completed and validated prior to product release. For example, "I am a user, and when I wave my hands beneath the dispenser within a range of 0–10 centimeters, soap is dispensed more than 95 percent of the time." Within the current (non-intersectional) paradigm, the user in this story is unmarked: their gender, race, age, class, and so on are not specified. If we shift to an intersectional framework, one of the implications is that we must restructure testing at all stages, from early prototypes through quality control in mass production, around what Algorithmic Justice League founder Joy Buolamwini has described as *intersectional benchmarks*.[89]

Retrofitting against Racism

In the long view, we live in the relatively early stages of a shift of these concerns from margins to center. Design justice is not yet a community of practice that is powerful enough to retool design processes writ large. For the moment, instead, each inequitable design outcome is read as an outlier or a quirk. For example, instances of obvious racism or sexism in algorithmic decision systems are framed as unfortunate byproducts of a system of technology design that is, overall, seen as laudable—and, furthermore, unstoppable. Biased tools and sociotechnical systems occasionally generate attention, typically through public outrage on social media followed by a few news stories. At that point, the responsible design team, institution, or firm allocates a small amount of resources to correct the flaw in what is seen as an otherwise excellent product. Yet there is a world of difference between post hoc debiasing of existing objects and systems, even if done to meet intersectional benchmarks, and the inclusion of design justice principles from the beginning. This is not to say that the former is never worthwhile. Our world is composed of a vast accretion of hundreds of years of designed objects, systems, and the built environment. Most have not been designed with the participation or consent of, let alone accountability to or control by, communities marginalized within the matrix of domination. Few have been designed or tested using an intersectional lens. In this context, "retrofitting against racism" is a key component of improving and equalizing life chances and experiences for subjects at disparate locations within the race/class/gender/disability matrix. That said, a successful paradigm shift would obviate the need to engage in post hoc fixes for designed objects and systems that constantly produce inequitable outcomes.

From Algorithmic Fairness to Algorithmic Justice: Color Blindness, Symmetrical Treatment, Individualization of Equality, and the Erasure of Historical Discrimination

One of the most urgent areas in which to apply design justice principles is algorithmic decision support systems. There is a growing awareness of algorithmic bias, both in popular discourse and in computer science as a field. An ever-growing body of journalism and scholarship demonstrates that algorithms unintentionally reproduce racial and/or gendered bias (less attention has been focused so far on algorithms and ableism, and questioning algorithmic reproduction of class inequality is barely on the table since financial risk calculation is so deeply normalized as to be hegemonic).[90] Algorithms are used as decision-making tools by powerholders in sectors as diverse as banking, housing, health, education, hiring, loans, social media, policing, the military, and more. Design justice calls for an analysis of how algorithm design both intentionally and unintentionally reproduces capitalism, white supremacy, patriarchy, heteronormativity, ableism, and settler colonialism.

For example, Safiya Noble, in her work *Algorithms of Oppression*, focuses our attention on the ways that search algorithms reproduce the matrix of domination through misrepresentation of marginalized subjects, especially through the circulation of hypersexualized images of Black girls and women (what Patricia Hill Collins calls *controlling images*).[91] Virginia Eubanks, in *Automating Inequality*, unpacks how algorithmic decision support systems that punish poor people were implemented as a right-wing strategy to limit and roll back hard-fought access to social welfare programs that were won by organized poor people's movements.[92] Kate Crawford and the AI Now Institute at NYU are producing a steady stream of critical work. For example, they ask us to consider what it would look like if search algorithms operated according to a logic of agonistic democracy,[93] and exhort us to imagine how algorithms might "acknowledge difference and dissent rather than a silently calculated public of assumed consensus and unchallenged values."[94] Joy Buolamwini, in her work with the Algorithmic Justice League, argues that we must develop intersectional training data, tests, and benchmarks for machine-learning systems.[95] Buolamwini is best known for demonstrating that facial analysis software performs worst on women with darker skin tones, but also advocates for greatly increased regulation and oversight of facial analysis tools, against their use by military or law enforcement, and fights to limit their use against marginalized people across areas as diverse as hiring, housing, and health care.[96]

There is a growing community of computer scientists focused specifically on challenging algorithmic bias. Beginning in 2014, the FAT* community emerged as a key hub for this strand of work.[97] FAT* has rapidly become the most prominent space for computer scientists to advance research about algorithmic bias: what it means, how to measure it, and how to reduce it. Papers about algorithmic bias are now regularly published in mainstream HCI journals and conferences. The keynote speech at the 2018 Strata Data Conference in Singapore focused on the need to use machine learning to monitor and counter algorithmic bias in machine-learning systems as they are deployed in myriad areas of life.[98] This is all important work, although the current norm of single-axis fairness audits should be replaced by a new norm of intersectional analysis. In some cases, this will require the development of new, more inclusive training and benchmarking data sets. At the same time, design justice as a framework also requires us to question the underlying set of assumptions about "inclusion," as STS scholar Os Keyes insists in their brilliant critique of the reproduction of the gender binary through data ontologies and algorithmic systems.[99] Design justice also involves a critique of the idea of "fairness" that nearly all these efforts contain, as Anna Lauren Hoffman reminds us in her recent paper on the limits of antidiscrimination discourse.[100]

As Patricia Hill Collins writes about the erasure of historical discrimination, the individualization of equality, and the concept of "symmetrical treatment" that characterize the ideology of "color blindness" in the post-*Brown v. Board of Education* US legal system: "Under this new rhetoric of color-blindness, equality means treating all individuals the same, regardless of differences they brought with them due to the effects of past discrimination or even discrimination in other venues."[101] What's more, the rhetoric of color blindness functions "as a new rule that maintains long-standing hierarchies of race, class, and gender while appearing to provide equal treatment."[102] Ruha Benjamin, in *Race After Technology*, develops the term "the New Jim Code" to highlight the ways that algorithmic decision systems based on historical data sets reinforce white supremacy and discrimination even as they are positioned by their designers as "fair," in the "colorblind" sense. Racial hierarchies can only be dismantled by actively antiracist systems design, not by pretending they don't exist.[103]

Unfortunately, most current efforts to ensure algorithmic fairness, accountability, and transparency operate according to the logic of individualized equality, symmetrical treatment, color blindness, and gender blindness. The operating assumption is that a fair algorithm is one that shows no group bias in the distribution of true positives, false positives, true negatives, and false negatives. For example, the widely read ProPublica article "Machine Bias" demonstrated that a recidivism risk algorithm overpredicted the likelihood of Black recidivism and underpredicted the rate of white recidivism.[104] The algorithm allocated more false positives to Black people and more false negatives to white people. A debate ensued about whether the algorithm was really biased, and if so, how it could be "fixed."

Design justice leads us to several key insights about this approach. First, the use of biased risk-assessment algorithms should not be discussed without reference to the context of the swollen prison industrial complex (PIC). A prison abolitionist stance does not support allocating additional resources to the development of tools that extend the PIC, even to make them "less biased." Instead, pretrial detention should be minimized as much as possible and ultimately eliminated.

Second, in areas where it does make sense to invest in attempts to monitor, reveal, and correct algorithmic bias, such efforts must be intersectional, rather than single-axis. For example, in bias audits, we need to know the false positive rates for white men, white women, Black men, and Black women.

Third, we should challenge the underlying assumption that our ultimate goal in algorithm design is symmetrical treatment. In other words, we have to raise the question of whether algorithm design should be structured according to the logic of "fairness," read as color and gender blindness, or according to the logic of racial, gender, and disability justice. The former implies that our goal is a fair algorithm that "treats all individuals the same," within the tightly bound limits of its operational domain and regardless of the effects of past or present-day discrimination. The latter implies something else: that the end goal is to provide access, opportunities, and improved life chances for all people, and that this requires redistributive action to undo the legacy of hundreds of years of discrimination and oppression. We need to discuss the difference between algorithmic colorblindness and algorithmic justice.

For example, consider an algorithm for university admissions. An (individualized) algorithmic fairness approach attempts to ensure that any two individuals with the same profile, but who differ only by, say, gender, receive the same recommendation (admit/waitlist/decline). Auditing an admissions algorithm under the assumptions of algorithmic fairness can be conducted through paired-test audits: submit a group of paired, identical applications, but change only the gender of one of the applicants in each pair and observe whether the system produces the same recommendation for each. If the algorithm recommends admission for more men than women (at a statistically significant level) in otherwise identical paired applications, we can say that it is biased against women. It needs to be retrained and reaudited to ensure that this bias is eliminated. This is the approach proposed by most of the researchers and practitioners working on algorithmic bias today.[105]

To modify this approach to be intersectional rather than single-axis may be more difficult but does not require a fundamental shift. An intersectional paired-test audit of the admissions algorithm requires submitting a far greater number of paired applications, with groups of applications that are nearly identical but with modified identity markers across multiple axes of interest within the matrix of domination: race, class, gender identity, sexual orientation, disability, and citizenship status, for example. This allows analysis of whether the system is biased against, say, Black Disabled men, queer noncitizen women, and so on. One question about this approach is how many identity variables to include because each adds complexity (and, in many situations, time and cost) to the audit. However, most of the researchers, developers, and engineers who are interested in correcting for algorithmic bias can probably be convinced that algorithmic bias analysis and correction should be intersectional and should at least include categories typically protected under US antidiscrimination law, such as sex, race, national origin, religion, and disability.

Now, imagine auditing the same admissions algorithm, but under the assumptions of algorithmic justice. This approach is concerned not only with individualized symmetrical treatment, but also with the individual and group-level effects of historical and ongoing oppression and injustice within the matrix of domination, as well as how to ultimately produce a more just distribution of benefits, opportunities, and harms across all groups of people. In our example, this means that the algorithm designers must discuss, debate, and decide upon what they believe to be a just distribution of outcomes. For instance, they might decide that a just allocation of admissions decisions would produce an incoming class with a gender distribution that mirrors the general population (about 51 percent women). They might further decide that they would seek an incoming class in which intersecting race, class, gender, and disability identities also mirrored the proportions in the general population. Alternately, they might decide that their goal was to correct, as rapidly as possible, the currently skewed distribution of enrolled students across all four undergraduate years. In this case, if the current student population greatly underrepresented, say, Latinx students in relation to their demographic proportion of the broader population, then the admissions algorithm would be calibrated to admit a greater proportion of Latinx first years to "make up for" underadmissions in previous years. To take the thought experiment much further, perhaps the algorithm developers would decide that the goal was to correct bias in the admissions demographic data across the full institutional lifetime. To correct for the systematic exclusion of women and people of color during the first hundred years of the university's existence, the algorithm might be calibrated to admit an entire class of women of color. Here, we are raising the question, "What would algorithmic reparations look like?"

My point is not to argue that this is exactly the outcome that should be sought for all algorithmic decision-making systems. My point is that the question of "What is a just outcome?" is not even on the table. Instead, our conversation remains tightly limited to a narrow, individualized conception of fairness. In the US context, it is highly unlikely that an algorithmic justice approach will advance, not least because in many instances this approach would violate existing antidiscrimination law. Nevertheless, as the conversation about algorithmic bias swells, and as we develop an array of tools to detect, mitigate, and counter algorithmic bias, we must propose alternate approaches, tools, configurations, and outcome metrics that would satisfy algorithmic justice. We must ask questions such as this: Within any decision-making system, what distribution of benefits do we believe is just?

Hard Coding Liberation: New Developments in Scholarship and Practice

This chapter began with a story about the (lack of) affordances of popular social media platforms for community organizing. It then opened into a critical discussion of the distribution of affordances under the matrix of domination, introduced the concept of disaffordances and dysaffordances, and described how affordances may be experienced as microaggressions. Design justice rethinks the universalizing assumptions behind affordance theory, requires us to ask questions about how inequality structures affordance perceptibility and availability, and takes both intentionally and unintentionally discriminatory design seriously.

Design justice builds on a long history of related approaches, such as value-sensitive design, universal design, and inclusive design. VSD provides some useful tools; however, it leaves many of the central questions of design justice unaddressed. VSD is descriptive rather than normative: it urges designers to be intentional about encoding values in designed systems but does not propose any particular set of values at all, let alone an intersectional understanding of racial, gender, disability, economic, environmental, and decolonial justice. VSD never questions the standpoint of the professional designer, doesn't call for community inclusion in the design process (let alone community accountability or control), and doesn't require an impact analysis of the distribution of material and symbolic benefits that are generated through design. Values are treated as disembodied abstractions, to be codified in libraries from which designers might draw to inform project requirements. In other words, in VSD we are meant to imagine that incorporating values into design can be accomplished largely by well-meaning expert designers. In design justice, by contrast, values stem from the lived experience of communities and individuals who exist at the intersection of systems of structural oppression and resistance.

The disability justice movement created many of the key concepts that underpin design justice, and has long articulated critiques of universalist design approaches. Much, or perhaps most, design work imagines itself to be universal: designers intend to create objects, places, or systems that can be used by anybody. Design justice challenges the underlying assumption that it is possible to design for all people. Instead, we must always recognize the specificity of which kinds of users will benefit most. Does this mean that design justice denies the very possibility of universal design? Perhaps design justice is an approach that can be applied to both universalist and inclusive (one size fits one) design projects. Design justice might help universalist design processes more closely approach their never fully realizable goals and provide useful insights to inclusive design processes. Retooling for design justice means developing new approaches to key design methods like A/B tests, benchmarks, user testing, and validation. In addition, this approach raises questions about the current dominant approach to the design of algorithmic decision support systems.

In the future, design justice must also help inform the development of emergent sociotechnical systems like artificial intelligence. Beyond *inclusion* and *fairness* in AI, we need to consider *justice*, *autonomy*, and *sovereignty*. For example, how does AI reproduce colonial ontology and epistemology? What would algorithmic decision making look like if it were designed to support, extend, and amplify indigenous knowledge and/or practices? In this direction, there is a growing set of scholars interested in decolonizing technologies, including AI systems. For example, designers Lewis, Arista, Pechawis, and Kite draw from Hawaiian, Cree, and Lakota knowledge to argue that indigenous epistemologies, which tend to emphasize relationality and "are much better at respectfully accommodating the non-human," should ground the development of AI.[106] Lilly Irani et al. have argued for the development of postcolonial computing;[107] Ramesh Srinivasan has asked us to consider indigenous database ontologies in his book *Whose Global Village*;[108] and anthropologist and development theorist Arturo Escobar has recently released a sweeping new book titled *Designs for the Pluriverse*.[109]

Escobar draws from decades of work with social movements led by indigenous and Afro-descended peoples in Latin America and the Caribbean to argue for *autonomous design*. He traces the ways that most design processes today are oriented toward the reproduction of the "one world" ontology. This means that technology is primarily used to extend capitalist patriarchal modernity and the aims of the market and/or the state, and to erase indigenous ways of being, knowing, and doing (ontologies, epistemologies, practices, and lifeworlds). Escobar argues for a decolonized approach to design that focuses on collaborative and place-based practices and that acknowledges the interdependence of all people, beings, and the earth. He insists on attention to what he calls the ontological dimension of design: all design reproduces certain ways of being, knowing, and doing. He's interested in the Zapatista concept of creating "a world where many worlds fit,"[110] rather than the "one world" project of neoliberal globalization.

Happily, research centers, think tanks, and initiatives that focus on questions of justice, fairness, bias, discrimination, and even decolonization of data, algorithmic decision support systems, and computing systems are now popping up like mushrooms all around the world. As I mentioned in this book's introduction, these include Data & Society, the AI Now Institute, and the Digital Equity Lab in New York City; the new Data Justice Lab in Cardiff; and the Public Data Lab.[111] Coding Rights, led by hacker, lawyer, and feminist Joana Varon, works across Latin America to make complex issues of data and human rights much more accessible for broader publics, engage in policy debates, and help produce consent culture for the digital environment. They do this through projects like *Chupadados* ("the data sucker").[112] Others groups include Fair Algorithms, the Data Active group, the Center for Civic Media at MIT; the Digital Justice Lab, recently launched by Nasma Ahmed in Toronto; Building Consentful Tech, by the design studio And Also Too in Toronto; the Our Data Bodies Project; and the FemTechNet network.[113] There is also a growing number of conferences and convenings dedicated to related themes; besides FAT*, 2018 saw the Data4BlackLives conference, the 2018 Data Justice Conference in Cardiff, and the AI and Inclusion conference in Rio de Janeiro, organized by the Berkman-Klein Center for Internet & Society, ITS Rio, and the Network of Centers; as well as the third design justice track at the Allied Media Conference in Detroit.[114]

Regardless of the design domain, design justice explicitly urges designers to adopt social justice values, to work against the unequal distribution of design's benefits and burdens, and to attempt to understand and counter white supremacy, cisheteropatriarchy, capitalism, ableism, and settler colonialism, or what Black feminist thought terms the *matrix of domination*. Design justice is interested in how to hardcode the liberatory values of intersectional feminism at every level of designed objects and systems, including the interface, the database, the algorithm, and sociotechnical practices "in the wild." What's more, this approach is interested not only in designed objects and systems, but in all stages of design, from the framing and scoping of design problems (chapter 3) to designing and evaluating particular affordances (as we explored in this chapter) to the sites where we do design work (chapter 4). The next chapter (chapter 2) unpacks the implications of design justice for the question, "Who gets to be a designer?"

2 Design Practices: "Nothing about Us without Us"

Figure 2.1
Cover illustration for "Nothing About Us Without Us: Developing Innovative Technologies For, By and With Disabled Persons" by David Werner, 1998, http://www.dinf.ne.jp/doc/english/global/david/dwe001/dwe00101.html.

> Today the tech industry does not look like America, and that has a significant influence on the types of products and services that get created ... When the lived experience of underrepresented communities is omitted from the product development cycle, the usefulness of the technology becomes biased towards one group.
>
> —Kapor Capital Founders' Commitment, 2015

> If you have come here to help me, you are wasting your time. But if you have come because your liberation is bound up with mine, then let us work together.
>
> —Lilla Watson, Australian Aboriginal activist and artist

In August 2017, a software engineer at Google ignited a firestorm of controversy with a memo titled "Google's Ideological Echo Chamber," which was widely circulated inside the company before it spread worldwide via social media and, later, mainstream news outlets. The author argued that underlying biological differences between men and women, rather than sexism, may explain the underrepresentation of women in software development and in high-level positions inside technology companies, as well as gendered salary differences, and that programs at Google that are designed to increase diversity and support women actually discriminate against men.

Response was swift. Within days, Google Chief Executive Sundar Pichai issued a public statement condemning the memo, and the employee who wrote the memo was fired.[1] The memo's author then sued the company for discrimination against whites, Asians, and men. The memo and lawsuit were hotly debated on social media, in blogs, and by scientists; dozens, if not hundreds, of news stories, op-eds, and think pieces were written about the case. Many provided detailed refutation of the memo's arguments; others noted that there is some support for some of the memo's scientific claims, but not for the author's conclusions about diversity policies; still others attempted to summarize arguments both for and against the memo's claims.[2] Some writers excoriated the culture of technology companies that allows misogyny and racism to flourish,[3] or suggested strategies for Silicon Valley firms to create more diverse and inclusive working environments.[4] Still others drew attention to women's many contributions to the field of software development,[5] from Ada Lovelace, the first software developer, to Grace Hopper, creator of one of the first compilers, to Katherine Johnson, the

Black woman whose calculation of space flight trajectories contributed to NASA's first moon landing (a story adapted for the big screen in the Hollywood film *Hidden Figures*).[6]

The infamous Google memo was only one moment in a series of increasingly high-profile controversies about sexism, racism, sexual harassment, and rape culture in the tech sector. These have had growing impact, in no small part due to the explosion of the #MeToo movement. Yet despite publicly repudiating the ideas in the Echo Chamber memo, tech companies continue to be sites for the systematic reproduction of the matrix of domination. Tech companies reproduce intersectional oppression through their hiring, retention, and promotion practices; through internal corporate culture that tolerates misogyny, racism, and sexual harassment; and through the products they design. For example, even as PR teams clambered over one another to publicly repudiate the misogyny inherent in the Echo Chamber memo's arguments, companies like IBM were meeting with the Trump administration to discuss bids on a government contract to build a "good immigrant/bad immigrant" prediction system.[7] Courts across the country were signing contracts to implement recidivism risk prediction software that has been shown to be racially biased.[8] Data about millions of low-income women was being ingested and analyzed by black box algorithms to determine whether these women would receive or be cut off from public benefits.[9]

The story of the Google memo, and of the pushback against it, illustrates three key points. First, racism and sexism (or, to describe oppression in structural terms that are more difficult to individualize, white supremacy and heteropatriarchy) remain pervasive within the culture of the most powerful technology companies in the world. The Google memo was not only notable because it was written by a software engineer at Google, but also because of its widespread circulation within the company and its sympathetic reception from many of the author's colleagues.

Second, although these ideas remain pervasive and they continue to structure practices in many spheres of life, they are no longer considered socially acceptable. Those who consciously hold these ideas understand this and exploit it to their advantage. The Echo Chamber memo is only the latest work in a long-standing genre that attempts to position misogynist ideas as reasonable arguments that are unfairly suppressed and marginalized due to "political correctness." This narrative of white (cis)male marginalization and oppression, on the defense against irrational attacks by feminists and/or people of color, is deeply embedded within the larger political climate. White male "marginalization" has long been a core narrative strategy of the right wing in the United States.[10] Indeed, the victorious Trump campaign tapped this wellspring of fears about the erosion of white masculinity, along with deeply rooted narratives of white womanhood under threat from Black and Latino men, to secure the world's most powerful elected position, with 62 percent of white men's and 53 percent of white women's votes, in the 2016 general election.[11]

Third (and most salient here), prominent critiques of the Google memo, like most stories about sexism and racism in Silicon Valley, are typically framed in terms of the untapped capacity of women, Black people, Indigenous people, and/or people of color (B/I/PoC), to perform well in jobs currently dominated by white and Asian cisgender men. Many laud the benefits of "diverse teams" for capitalist profitability. Sexist and racist discourse and practice within the technology industry are nearly always delinked from broader and deeper critiques of the ways that tech reproduces white supremacy, heteropatriarchy, capitalism, and settler colonialism—not only through employment practices, but through all aspects of technology design. Although the matrix of domination shapes a range of tech sector activities, including (but not limited to) employment, choice of intended users, scoping, affordances, access to capital, platform ownership and governance, and more, the conversation about challenges (and solutions) usually remains within an "employment diversity" frame.

Employment diversity is important. However, ultimately, design justice challenges us to push beyond the demand for more equitable allocation of professional design jobs. Employment diversity is a necessary first move, but it is not the far horizon of collective liberation and ecological sustainability. The goal of this book is to spur our imaginations about how to move beyond a system of technology design largely organized around the reproduction of the matrix of domination. In its place, we need to imagine how all aspects of design can be reorganized around human capabilities, collective liberation, and ecological sustainability.

Designers: Who Gets (Paid) to Do Design?

To begin with, design justice as a framework recognizes the universality of design as a human activity. As noted in the introduction, *design* means to make a mark, make a plan, or problem-solve; all human beings thus participate in design.[12] However, though all humans design, not everyone gets paid to do so. Intersectional inequality systematically structures paid professional design work. Professional design jobs in nearly all fields are disproportionately allocated to people who occupy highly privileged locations within the matrix of domination. At the same time, the numerous expert designers and technologists who are not wealthy and/or educationally privileged white cis men have often been ignored, their labor appropriated, and their stories erased from the history of technology.[13] Also, professional designers constantly draw both from one another and from the unsung design work of everyday people. Although the discussion that follows could easily apply to any professionalized design field, I will focus on the software and technology industries. Designers in this sector are highly rewarded, both economically and culturally, and have achieved status as iconic figures who stand in for the promise of innovation and entrepreneurialism under informational capitalism.

In recent years, there has been a growing public conversation about the fact that the most advanced sector of the economy might well be the most unequal. In 2016, several technology firms, under pressure from mobilized publics, released diversity data about their employment practices. Unsurprisingly, this data does not paint a flattering picture of progress toward gender and racial equality in the tech sector. White and Asian cis men dominate technology jobs. For example, in the United States, women overall hold 26 percent of tech jobs, Black women hold just 3 percent of computer programming jobs, and Latinas hold 2 percent.[14] As feminist media anthropologist Christina Dunbar-Hester notes, gender disparity in the software industry is far worse within the supposedly "open" arena of free/libre and open-source software (F/LOSS): just 2 percent of F/LOSS developers are women, compared to 30 percent of developers who work on proprietary software.[15] A 2016 report by Intel found that nearly two-thirds of tech workers are white.[16] Sector-wide employment trends are not steadily advancing toward increasing diversity; instead, women and/or B/I/PoC sometimes gain ground, sometimes lose ground.[17]

Even when women and/or B/I/PoC are employed in technology design, development, and product management, only a handful have positions at the top of these extremely hierarchical organizations. Gender diversity on the boards of top tech companies tends to range from just 10 percent to 25 percent (almost exclusively white) women. For example, Apple's board currently has six men and two women, Google (Alphabet) has nine and two, Facebook seven and two, and so on. Yahoo, with a board composed of six men and three women, is the top-tier tech firm that comes closest to gender parity at the highest decision-making level.[18]

Extreme gender disparity in computing was not always the norm. Indeed, *computers* originally were human beings, often women, who performed extensive calculations in fields including astronomy, ballistics, economic analysis, and more.[19] The world's first professional computer programmers were arguably the six women (Fran Bilas, Betty Jean Jennings, Ruth Lichterman, Kay McNulty, Betty Snyder, and Marlyn Wescoff) tasked with programming ENIAC to calculate ballistics trajectories during World War II.[20] In the early days of modern computer science, women made up a much higher proportion of computer scientists. One study from Google estimated that in the 1980s, 37 percent of computer science majors were women; by 2012, the proportion had dropped to 18 percent.[21] Dunbar-Hester discusses some of the reasons for this shift. As a skillset and occupational path, computer programming was initially marginal, unprestigious, poorly understood, and (crucially) not particularly well-paid. As programming took center stage in the new information economy, men pushed women to the side.[22] White male geek culture, replete with heteropatriarchal cultural structures, forms of humor, and mechanisms for normalizing white cis male standpoints, came to rule the roost.[23]

Dismal equity statistics reflect broader raced and gendered patterns that persist across nearly all sectors of the economy. Racial and gender inequality in who gets paid to do design work is also shaped by educational access inequalities,[24] and I will return to questions about diversity in technology education, as well as the purpose of such education, in chapter 5, which focuses on the pedagogy of design justice. For now, it is enough to say that the many "diversity in tech" initiatives are important. If they are matched by systematic shifts in hiring, mentorship, and retention practices across the technology industry, such as those indicated by recent research into best practices in corporate gender parity efforts,[25] there may well be a long-term shift in trends toward more equitable employment in this industry. However, full parity remains unlikely without macro policy shifts to support systems-wide equalizers such as universal family policies, job protection, generous paid parental leave, child care services, and other key factors that are known to support gender-diverse employment, retention, and career advancement.[26]

Diversity Is Good for Capitalist Profitability

Although employee diversity is certainly a laudable goal, it remains comfortably within the discourse of (neo)liberal multiculturalism and entrepreneurial citizenship.[27] Indeed, there is a growing managerial literature on the competitive business advantages of employee diversity. Diverse firms and product teams have repeatedly been shown to make better decisions, come up with more competitive products, and better understand potential customers. Racial and gender diversity are linked to increased sales revenue, more customers, and greater relative profits,[28] although some research complicates this narrative.[29] This is now fairly well understood in mainstream business literature. As the 2017 *Breaking the Mold* report notes: "McKinsey & Company has reported that companies in the top quartile in terms of racial diversity are 35 percent more likely to have financial returns higher than the national median in their industry. This research complements multiple studies which conclude that gender diversity clearly improves corporate financial performance."[30] The cited McKinsey report analyzed private data sets on employment diversity from 366 firms and found that "in the United States, there is a linear relationship between racial and ethnic diversity and better financial performance: for every 10 percent increase in racial and ethnic diversity on the senior-executive team, earnings before interest and taxes (EBIT) rise 0.8 percent."[31] Despite a growing body of studies that demonstrate at least correlation (if not causation) between employee diversity and capitalist profitability, as well as shifting mainstream cultural norms that favor increased gender and racial/ethnic diversity, corporate leadership remains dominated by white cis men across all sectors of the economy. Women make up just 16 percent of executive teams in US companies, 12 percent in the United Kingdom, and 6 percent in Brazil. In terms of race, 97 percent of US companies have senior leadership teams composed primarily of white people.[32]

In other words, under the informational stage of racial capitalism, employee diversity is seen by most of the managerial class as an input to increased efficiency, innovation, market domination, and capital accumulation. However, despite steadily increasing interest in establishing a diverse pool of designers, developers, product managers, and other tech workers, the industry persistently fails to meaningfully diversify. What's more, structural inequality is rarely mentioned, let alone challenged. Because design justice as a framework includes a call to dismantle the matrix of domination and challenge intersectional, structural inequality, it requires more than a recognition that employment diversity increases capitalist profitability. Employment in paid design fields is important, but is not the whole picture. Design justice also involves rethinking other aspects of design practice, including the intended design beneficiaries: the "users."

Imagined Users: Whose Tech?

For whom do we, as a society, design technology? Journalist and feminist activist Laurie Penny puts it this way:

> There is nothing wrong with making things that people want. The problem is that personhood and desire are constrained by capital; money affects whose wants appear to matter. The kids in Startup House may want a pizza delivery drone, but not in the same way low-income families want health care, or the elderly men lying in their own faeces on Howard Street want a safe place to sleep. There is nothing wrong with making things people want. It's just that too little attention is being paid to the things people need. The wants and needs of young, healthy, middle-class people with connections and a reasonable amount of spare cash are overrepresented among Start-up City's priorities. For one thing, those are the problems with solutions that sell. For another, given a few million dollars and a team of semi-geniuses, those problems are easy to solve. Structural social injustice and systemic racism are harder to tackle.[33]

Penny's critique of classed user prioritization within capitalist start-up scenes can be extended: default imagined users are raced, classed, and gendered within a worldview produced by the matrix of domination, internalized and reproduced within technology design teams. Designers most frequently assume that the unmarked user has access to several very powerful privileges, such as US citizenship, English language proficiency, access to broadband internet, a smartphone, a normatively abled body, and so on.

User-Centered Design, the "Unmarked" User, and the Spiral of Exclusion

User-centered design (UCD) refers to a design process that is "based upon an explicit understanding of users, tasks, and environments; is driven and refined by user-centered evaluation; and addresses the whole user experience. The process involves users throughout the design and development process and it is iterative."[34] Over time, UCD has become the recommended design approach within many firms, government bodies, and other institutions. However, UCD faces a paradox: it prioritizes "real-world users." Yet if, for broader reasons of structural inequality, the universe of real-world users falls within a limited range compared to the full breadth of *potential* users, then UCD reproduces exclusion by centering their needs. Put another way, design always involves centering the desires and needs of some users over others. The choice of *which users* are at the center of any given UCD process is political, and it produces outcomes (designed interfaces, products, processes) that are better for some people than others (sometimes very much better, sometimes only marginally so). This is not in and of itself a problem. The problem is that, too often, this choice is not made explicit.

In addition, designers tend to unconsciously default to imagined users whose experiences are similar to their own.[35] This means that users are most often assumed to be members of the dominant, and hence "unmarked" group: in the United States, this means (cis) male, white, heterosexual, "able-bodied," literate, college educated, not a young child and not elderly, with broadband internet access, with a smartphone, and so on. Most technology product design ends up focused on this relatively small, but potentially highly profitable, subset of humanity. Unfortunately, this produces a spiral of exclusion as design industries center the most socially and economically powerful users, while other users are systematically excluded on multiple levels:

their user stories, preferred platforms, aesthetics, language, and so on are not taken into consideration. This in turn makes them less likely to use the designed product or service. Because they are not among the users, or are only marginally present, their needs, desires, and potential contributions will continue to be ignored, sidelined, or deprioritized.

It is tempting to hope that employment diversity initiatives in the tech sector, if successful over time, will solve this problem. Diversifying the technology workforce, as noted above, is a good move, but unfortunately, it will not automatically produce a more diverse default imagined user. Research shows that unless the gender identity, sexual orientation, race/ethnicity, age, nationality, language, immigration status, and other aspects of user identity are explicitly specified, even diverse design teams tend to default to imagined users who belong to the dominant social group.[36]

There is growing awareness of this problem, and several initiatives attempt to address it through intentional focus on designing together with communities that are usually invisibilized. For example, the Trans*H4CK series of hackathons focuses on trans* and gender-nonconforming communities. Contratados.org[37] is a site built by the Center for Migrant Rights that operates like Yelp, but for migrant workers, to let them review potential employers and recruitment agents, educate them about their rights, and protect them from transnational recruitment scams.[38] Such efforts to design together with users from communities that are mostly overlooked by design industries are important. However, they remain small-scale. What's more, individual inclusive design projects cannot, on their own, transform the deeply entrenched systemic factors that militate toward design that constantly centers an extremely limited set of imagined users.

Design Justice and Lead User Innovation

Another way to think about the relationship among users, design processes, and the matrix of domination is through MIT management professor Eric Von Hippel's concepts of *lead user innovation, information asymmetry between manufacturers and users,* and *variance in user product needs*. Design justice focuses on the ways that race, class, gender, and disability structure both information asymmetries and variance in user product needs.

In Von Hippel's widely influential text *Democratizing Innovation*, he demonstrates that a great deal of technological innovation—perhaps the majority—is actually conducted by what he terms *lead users*. Von Hippel uses a powerful mix of case studies, economic theory, and industry data to demonstrate how this process works across a range of sectors, from extreme sports to software development. He identifies several underlying shared principles that help explain why users innovate and modify commercially available products, why users often freely share these innovations with one another, and why firms frequently fail to develop products that meet user needs. For example, Von Hippel demonstrates that information asymmetry between manufacturers and users is one of the underlying forces that supports lead user innovation. In a nutshell, when the cost (in time and energy) of communicating a specific kind of user need to the manufacturer is high, it often makes more sense for users to modify products on their own than to attempt to convince manufacturers to do so.[39] In addition, Von Hippel shows that it is highly likely that certain groups of users—in particular, those who are too few to ensure that the manufacturer will benefit from economies of scale—will be more likely than others to have unmet needs. Because the group of users who push the limits of any particular technology tends to overlap closely with those who are the most skilled in the activity the technology is meant to support (the lead users), there is a permanent tension between economies of scale for manufacturers and the most likely location of innovation. To mitigate this problem, Von Hippel suggests strategies for firms that hope to learn from and more effectively incorporate lead user innovation into their product development cycles.

However, Von Hippel's otherwise compelling theory of lead user innovation does not engage with race, gender, class, or other axes of structural inequality. For example, he never considers the implications of information asymmetry between manufacturing firms and users in a context in which a firm is controlled by, say, white men, but its users are more diverse. As we have seen, design, engineering, and decision making in firms are led by people from the dominant social groups, and so product specifications are likely to center the needs of people who belong to those groups. If white cis male designers, engineers, and decision-makers run most product design processes, white cis men will be more likely to have their needs met than members of other groups. The costs of communicating specific user needs will generally be higher for users from disadvantaged locations within the matrix of domination. Building on Von Hippel's theory, there is less information asymmetry between designers and users who occupy similar positions in the matrix of domination than between designers and users in very different locations.

However, even if design teams perfectly mirrored users in terms of standpoint within the matrix of domination, and even if the unequal costs of communicating specific user needs to decision-makers were addressed, firms would still face pressures from economies of scale to produce solutions optimized for the specifications of the most profitable group of users. As Von Hippel describes, because of economies of scale, firms have very strong incentives to foist existing solutions on all users, even where some users have different specifications. User product specifications for groups of users who are numerically a minority and/or whose purchasing power is relatively small are less likely to be met. Because purchasing power under white supremacist capitalist heteropatriarchy is unequally structured by race, class, and gender, product design ends up disproportionally prioritizing the user specifications of relatively wealthy white men.

Finally, Von Hippel never specifically explores how what he refers to as *variance in user product specifications* might be structured by race, class, gender identity, sexual orientation, and/or disability. For example, consider the gender identity and sexual orientation options on the popular dating site Tinder. In response to pressure from users with gender identities other than "man" or "woman," in 2016 the site began to allow selection of additional gender identities for display, such as trans*, nonbinary, and so on.[40] However, these new options only affect the displayed label in the gender field; they are not useful for the actual searching and matching function of the site. In the settings that determine who will see the user's profile, the options remain constrained to Man and Woman. Similarly, in the setting that determines which subset of profiles the user will see, there is no option to specifically see the profiles of users who have selected trans*, nonbinary, or other gender identity labels. In other words, the change is primarily cosmetic; it does not meet the variance in user product specification of interest to many trans*, GNC, or nonbinary users.[41]

"Stand-in Strategies" to Represent Communities That Are Not Really Included in the Design Process

Well-meaning designers and technologists often agree that including "diverse" end users in the design process is the ideal. However, many feel that this is usually, sometimes, or mostly impossible to realize in practice. To mitigate the potential problems that come from having no one with lived experience of the design problem actually participate in the design team, researchers and designers have suggested several strategies. Unfortunately, most of these strategies involve creating abstractions about communities that are not really at the table in the design process. Such strategies include design ethnography, focus groups, and a great deal of what passes for participatory design. Here I explore the most widely used "stand-in" strategy: user personas.

User Personas *User personas* are short, fictional characterizations of product users, often with a name, an image, and a brief description. They are widely used to guide a range of design processes, including UX and UI, graphic design, product development, architecture, service design, and more.[42] User personas are so widespread that there is even a small sector of firms in the business of providing tools for design teams to generate, manage, and share them. For example, the Userforge website (figure 2.2) allows rapid random generation of user personas and promises to help design teams "build empathy and develop focus quickly. Create realistic representations of your user groups in far less clicks than it would take using design software or word processors, which means you can start prioritizing design decisions and get to the wins sooner."[43]

User personas can be useful tools for communicating project goals, both within teams and firms and to other actors, including funders, investors, the press, and potential users. There is some evidence that user personas help designers stay focused on the intended use case.[44] In addition, some case-control studies have sought to demonstrate the utility of user personas for better design outcomes.[45] If they are developed in ways that are truly grounded in the lived experience of the community of end users, through careful research or by community

Build empathy and develop focus quickly.

Create realistic representations of your user groups in far less clicks than it would take using design software or word processors, which means you can start prioritizing design decisions and get to the wins sooner.

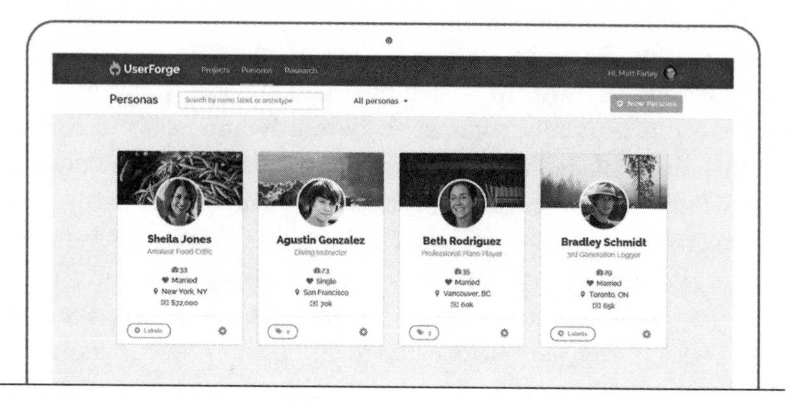

Figure 2.2
Userforge user persona generator. Screenshot from Userforge.com.

members themselves, they may be especially worthwhile. However, there is no systematic study that I was able to locate that examines whether the use of diverse user personas produces less discriminatory design outcomes.

Too often, design teams only include "diverse" user personas at the beginning of their process, to inform ideation. Occasionally, diverse user stories or personas are incorporated into other stages of the design process, including user acceptance testing. However, even if the design team imagines diverse users, creates user personas based on real-world people, and incorporates them throughout the design process, the team's mental model of the system they are building will inevitably be quite different from the user's model. Don Norman, one of the most important figures in User Centered Design (UCD), notes that in UCD "the designer expects the user's model to be identical to the design model. But the designer does not talk directly with the user—all communication takes place through the system image."[46]

To make matters worse, far too often user personas are created out of thin air by members of the design team (if not autogenerated by a service like Userforge), based on their own assumptions or stereotypes about groups of people who might occupy a very different location in the matrix of domination. When this happens, user personas are literally objectified assumptions about end users. In the worst case, these objectified assumptions then guide product development to fit stereotyped but unvalidated user needs. Sometimes, they may also help designers *believe* they are engaged in an inclusive design process, when in reality the personas are representations of designers' unvalidated beliefs about marginalized or oppressed communities. Unsurprisingly, there are no studies that compare this approach to actually including diverse users on the design team.

Disability Simulation Is Discredited; Lived Experience Is Nontransferable

There are several reasons that designers who attempt to preempt discriminatory design by imagining themselves as various kinds of users often fail. Ultimately, pretending to be another kind of person is not a good solution for design teams that want to minimize discriminatory design outcomes. Design theorist D. E. Wittkower argues against informal attempts to imagine diverse user experiences and in favor of *systematic phenomenological variation*.[47] Other design techniques to imagine diverse users include designation of a team member as a "user diversity advocate," diverse user personas, real-world user testing, formal audits, and iterative feedback and redesign cycles with real-world users after product launch, among others. None of these techniques are as good as the inclusion of diverse users on the design team throughout the process. For example, the supposedly beneficial design practice of "disability simulation" has been discredited by a recent meta-analysis.[48] In disability simulation, "a nondisabled person is asked to navigate an environment in a wheelchair in order, supposedly, to gain a better understanding of the experiences of disabled persons. These 'simulations' produce an unrealistic understanding of the life experience of disability for a number of reasons: the nondisabled person does not have the alternate skill sets developed by [Disabled people], and thus overestimates the loss of function which disability presents, and is furthermore likely to think of able-normative solutions rather than solutions more attuned to a [Disabled person's] life experience."[49] For example, abled designers typically focus on an ableist approach to technologically modifying or augmenting the individual bodies of Disabled people to approximate normative mobility styles, compared to Disabled people, who may be more interested in architectural and infrastructural changes that fit their own mobility needs. As Wittkower says, ultimately, attempting to imagine other people's experience is "no substitute for robust engagement with marginalized users and user communities. ... [systematic variation techniques], although worth pursuing, are strongly limited by the difficulty of anticipating and understanding the lived experiences of others."[50] A design justice approach goes further still: beyond "robust engagement," design teams should be led by and/or in other ways be formally accountable to marginalized users.

If You're Not at the Table, You're on the Menu Design justice does not focus on developing systems to abstract the knowledge, wisdom, and lived experience of community members who are supposed to be the end users of a product. Instead, design justice practitioners focus on trying to ensure that community members are actually included in meaningful ways throughout the design process. Another way to put this is "If you're not at the table, you're on the menu."[51] Design justice practitioners flip the "problem" of how to ensure community participation in a design process on its head to ask instead how design can best be used as a tool to amplify, support, and extend existing community-based processes. This means a willingness to bring design skills to community-defined projects, rather than seeking community participation or buy-in to externally defined projects. Ideally, design justice practitioners don't focus on how to provide incentives that we can dangle to entice community members to participate in a design process that we have already determined and that we control. Instead, design justice compels us to begin by listening to community organizers, learning what they are working on, and asking what the most useful focus of design efforts would be. In this way, design processes can be community-led, rather than designer- or funder-led. Another way to put this might be: "Don't start by building a new table; start by coming to the table."

Design Process: From Participation to Accountability to Ownership

This chapter began with a critique of the raced, classed, and gendered nature of employment in the technology sector, as well as the ways that the matrix of domination structures imagined users and focuses designer's imaginations about who we are designing for. Besides diversity in professional design jobs, design justice requires full inclusion in the design process of people with direct lived experience of the conditions the design team wants to change. What's more, in addition to equity (we need more diverse designers, and more diverse imagined users), design justice also emphasizes accountability (those most affected by the outcomes should lead design processes) and ownership (communities should materially own design processes and their outputs).

Participatory Design

The proposal to include end-users in the design process has a long history. The "participatory turn" in technology design, or at least the idea that design teams cannot operate in isolation from end users, has become increasingly popular over time in many subfields of design theory and practice. These include participatory design (PD), user-led innovation, user-centered design (UCD), human-centered design (HCD), inclusive design, and codesign, among a growing list of terms and acronyms.[52] Some of these approaches have been adopted by multinational technology companies. Top firms have recently created toolkits and methods to address inclusion in design. For example, in 2017, in a story for *Fast Company* about Airbnb's new inclusive design toolkit,[53] technology journalist Meg Miller writes: "Microsoft has an inclusive design kit and a general design strategy centered around the philosophy that designing for the most vulnerable among us will result in better products and experiences for all. Google focuses on accessibility practices for their developers for the same reasons. Industry leaders like John Maeda and Kat Holmes have built their careers on speaking on the importance of diversity in the field, and how human-centered design should encompass potential users of all different races, genders, and abilities."[54] Only some of these approaches and practitioners, however, ask key questions about how to do design work in ways that truly respond to, are led by, and ultimately benefit the communities most targeted by intersectional structural inequality.

The question of community accountability and control in supposedly inclusive design processes has recently come to the fore in public conversations about civic tech. Daniel X. O'Neil, one of the key early actors in the field, has written a blistering critique of civic tech's lack of community accountability or connection to existing social movements.[55] Artist, educator, and community technologist Laurenellen McCann calls for technologists to "build with, not for."[56] Both find fault with civic tech's frequent solutionism, disconnection from real-world community needs, and tech-centric ideas about how to address difficult social problems, as well as for ongoing reproduction of white cis male "tech bro" culture that alienates women, trans* folks, B/I/PoC, Disabled people, and other marginalized communities.[57] This debate is the latest incarnation of a long-standing conversation about the relationship between communities and technology development that has animated shifts in theory, practice, and pedagogy across fields including design, software development, science and technology studies, international development, and many others over the years.

For example, as early as the 1960s, in parallel with the rise of the Non-Aligned Movement (formerly colonized countries across the Global South that hoped to chart a path away from dependency on either the United States or the USSR),[58] the *appropriate technology movement* argued that technology should be cheap, simple to maintain and repair, small-scale, compatible with human creativity, and environmentally sustainable.[59] Writings by economist E. F. Schumacher[60] and popular manuals such as Stewart Brand's *Whole Earth Catalog*[61] focused attention on small, local economies powered by appropriate technology, and countercultural movements throughout the 1960s spawned thousands of organizations dedicated to locally governed, environmentally sustainable technologies that could be adapted to the contexts within which they were embedded, in opposition to one-size-fits-all megaprojects championed by both Cold War powers as keys to "international development."[62]

In Scandinavia, the field of participatory design (PD) was created by trade unionists working with software developers such as Kristen Nygaard. They hoped to redesign industrial processes, software interfaces, and workplace decision-making structures.[63] In PD, end users are included throughout. Philosopher of science, technology, and media Peter Asaro describes PD as "an approach to engineering technological systems that seeks to improve them by including future users in the design process. It is motivated primarily by an interest in empowering users, but also by a concern to build systems better suited to user needs."[64] Like many scholars, Asaro traces the roots of PD to the Norwegian Industrial Democracy Project (NIDP). In the 1960s, Scandinavian designers and researchers were concerned with the ways that the introduction of new technology in a workplace is often used to eliminate jobs, deskill workers, and otherwise benefit the interests of owners and managers over the interests of workers. The collective resources program of NIDP centered on bringing choices about technology into the collective bargaining process. According to Asaro, British researchers at the Tavistock Institute focused on a parallel strand of research about individual worker empowerment through technology design, known as *sociotechnical systems design*. Asaro also points to the UTOPIA project as the canonical first successful instance of PD. UTOPIA was a collaboration among the Nordic Graphic Workers Union, researchers, and technologists, who worked with newspaper typographers to develop a new layout application. UTOPIA was developed after earlier PD experiments had failed, in part because of the creative limitations of existing technologies.

For decades, software developers employing PD have met at the biannual Participatory Design Conference.[65] PD has been widely influential and has spread to fields such as architecture and urban planning,[66] computer software,[67] public services, communications infrastructure, and geographic information systems,[68] among others. The Nordic approach to PD is also characterized by an emphasis on the normative value of democratic decision making in the larger technological transformation of work, not only the microlevel pragmatic benefits of improved user interface design. However, in the US context, this broader concern is often lost in translation. Here, PD has sometimes (at worst) been reduced to an extractive process to gather new product ideas.[69]

From the 1980s through the early 2000s, a parallel set of concepts was developed by scholars such as Eric Von Hippel, whose studies of lead user innovation demonstrated that the vast majority of innovation in any given technological field is performed not by governments or formal research and development branches of corporations, but by technology end users themselves.[70] This insight led to changes in product design approaches across a wide range of fields. Technology appropriation researchers such as Ron Eglash[71] and Bar, Weber, and Pisani[72] have shown that user practices of hacking, modifying, remixing, and otherwise making technologies work for their own ends are enacted quite commonly across diverse contexts. Whereas lead user innovation focuses on the hacks that people implement to make technologies serve their needs, and technology appropriation theory centers activities outside of formal product or service design processes, human-centered design emphasizes better understanding of everyday user needs and experiences in professional technology design and development.[73] By the 1990s, design consultancies such as IDEO emerged to champion (and capitalize on) this approach by selling HCD and design thinking as a service to multinational firms, governments, educators, and NGOs.[74] An extensive community of practitioners and scholars also clusters around the term *codesign,* often used as an umbrella that includes various approaches to PD and HCD. This approach is reflected in the journal *CoDesign,* in annual codesign conferences, and in the appearance of the concept across multiple fields.[75]

In the tech sector, *lean product development,* an approach that emphasizes early and frequent tests of product assumptions with real-world users, has largely replaced top-down "waterfall" design approaches as established best practice.[76] This shift has been increasingly influential in civic tech and government tech circles as well. Lean and HCD approaches to civic tech led to innovations such as 18F, a unit within the federal government's General Services Administration (GSA) that is focused on bringing software development best practices to government, as well as the Chicago User Testing group (CUTgroup), based on the experience of the Smart Chicago Collaborative and meant to promote the inclusion of end users in product design.[77] These approaches certainly increase end user input into key design decisions, but most of them have little to say about community accountability, ownership, profit sharing, or credit for innovation.

Power Dynamics and the Ladder of Participation

Power shapes participation in all design processes, including in PD, and the politics of participation are always intersectionally classed, gendered, and raced. Asaro outlines several challenges in PD projects: for one, it's not enough to have end users simply join design meetings. In a workplace context (or in any context), some users will feel they have more power than others. For example, workers participating in a PD meeting with managers at the table may not feel comfortable saying what they mean or sharing their full experience. The same may be the case in any PD process in which socially dominant group members are in the same room as marginalized folks, but without skilled facilitation. In addition, engineers and professional designers may control the "PD" process relatively easily, based on their "expert" knowledge. What's more, according to Asaro, gender inequality shapes participation in design processes: "In many work contexts, the positions traditionally occupied by women are often viewed as being of lower value by management and unions. This undervaluing of women's work easily overflows into inequalities of participation in design activities, especially when combined with social prejudices that view technological design as a masculine pursuit. Unless gender issues in the design process are recognized and dealt with, there exists a strong possibility of gender inequalities being built into the technology itself."[78] In the worst case, PD processes may actually normalize cultural violence through seemingly participatory processes. As design scholar and practitioner Ramesh Srinivasan says, "Foucault points out that cultural violence is perpetuated through seemingly inclusive systems, what one today might describe as liberal or neoliberal. These systems appear democratic, yet in practice they subordinate beliefs and practices not in line with those who manufacture discourse and manipulate media and technology systems to maintain their power and privilege."[79]

Participatory Design, Community Knowledge Extraction, and Non-extractive Design

Many design approaches that are supposedly more inclusive, participatory, and democratic actually serve an extractive function. Sometimes this is intentional, as in design workshops run by multinational corporations with potential end users, in which the goal is explicitly to generate ideas that will then be turned into products and sold back to consumers.[80] More frequently, the intentions of the designers are good. Well-meaning designers employ PD techniques for a wide range of reasons. For one thing, the process of working with community members is enjoyable. It feels good to elicit design ideas and possibilities from "nondesigners," it can be quite fun and engaging for everyone involved, and it can feel empowering for both design professionals and community members. Unfortunately, this does not change the fact that in most design processes, the bulk of the benefits end up going to the professional designers and their institutions. Products, patents, processes, credit, visibility, fame: the lion's share goes to the professional design firms and designers. Community members who participate in design processes too often end up providing the raw materials that are processed for value further up the chain. Design justice practitioners are working to rethink extractive design processes and to replace them with approaches that produce community ownership, profit, credit, and visibility.

Legal scholar Barbara L. Bezdek, theorizing what she terms *development justice*, notes: "Sherry Arnstein, writing in 1969 about citizen involvement in planning processes in the United States, at the height of American racial and economic tensions, described a typology of citizen participation arranged as a ladder with increasing degrees of decision-making clout ranging from low to high. The Arnstein rungs ascend from forms of 'window-dressing participation,' through cursory information exchange, to the highest levels of partnership in or control of decision-making."[81] Bezdek revisits the Arnstein rungs and rethinks the rules that govern public participation in urban economic redevelopment projects. She proposes a revised set of principles for civic engagement, and a series of actions toward development justice. Arnstein's ladder might also be useful to further articulate community participation in any design process.

Consider figure 2.3, in which the X axis represents the design phase (in this case, based on the widely used five-phase model from the Stanford d.school), and the Y axis represents the degree of participation by people from the communities most affected by the design project (following Arnstein's ladder). Each horizontal wavy line represents a (hypothetical) visual shorthand for how community participation unfolds across the life cycle of an individual design project. Put aside for the moment the fact that design does not really proceed along a linear path from phase to phase and that there are many, many different design process models.[82] In reality, phases have porous boundaries and are revisited multiple times during the project life cycle. The point is to encourage a more complex understanding of participation and

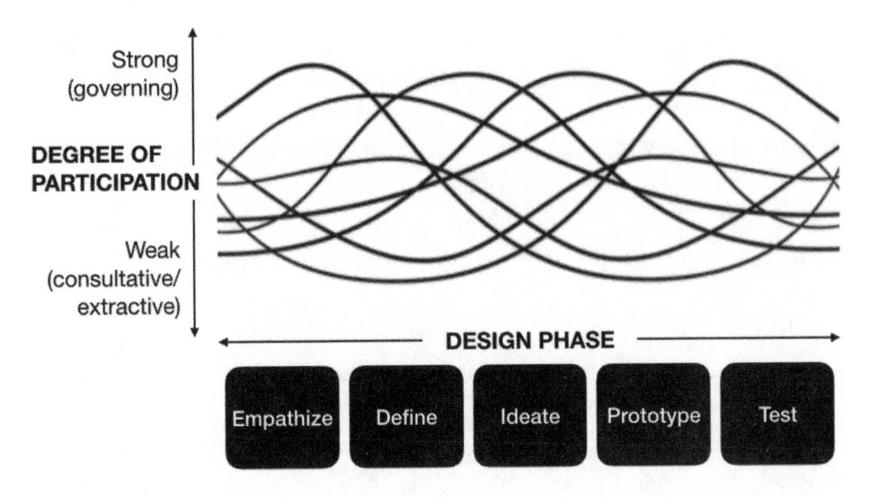

Figure 2.3
Analysis of community participation throughout the design process. *Source:* Author.

to emphasize that very few design processes are characterized by community control throughout. A version of this diagram may be a useful heuristic for thinking through questions of community participation, accountability, and control. A simple image that represents the participation waveform of a design project might be used in design criticism to analyze case studies, or it might be used by design justice practitioners to think through concrete community accountability and control mechanisms in projects that we work on.

Design Justice as Community Organizing

Design justice practitioners must also engage with fundamental questions about the definition of *community*. It is possible to criticize simplistic conceptions of community and representation without throwing up our hands and accepting the Thatcherite position that "there is no such thing as society."[83] The question of what a community is and how we can know what it wants is the domain of democratic theory and political philosophy. It is also a key question for fields including urban planning, participatory action research (PAR), development studies, and PD, among others.[84]

Design justice practitioners choose to work in solidarity with and amplify the power of community-based organizations. This is unlike many other approaches to PD, in which designers partner with a community but tend to retain power in the process: power to convene and structure the work, to make choices about who participates, and, usually, to make key decisions at each point. Analysis of political power in the design process—who sits at the table, who holds power over the project, what decision-making process is used—will be fundamental to the successful future articulation of design justice in theory and practice.

Ultimately, at its best, a design justice process is a form of community organizing. Design justice practitioners, like community organizers, approach the question of who gets to speak for the community from a community asset perspective.[85] This is rooted in the principle that wherever people face challenges, they are always already working to deal with those challenges; wherever a community is oppressed, they are always already developing strategies to resist oppression. This principle underpins what Black feminist author adrienne maree brown calls *emergent strategy*.[86] Emergent strategy grounds design justice practitioners' commitment to work with community-based organizations that are led by, and have strong accountability mechanisms to, people from marginalized communities. This contrasts with most other design approaches; even those that aim to involve users, citizens, or community members typically do so in a consultative process that ultimately is led by the professional designers. There are also many design approaches, including value sensitive design but also especially in urban planning, that seek "multi-stakeholder" participation. A typical design project focused on, say, gentrification and displacement might convene people representing as many of the different interests (stakeholders) as possible, such as long-time residents facing displacement; new, wealthier residents seeking housing; landlords; developers; planners; city officials; and so on. In contrast, a design justice project might engage with all these kinds of actors in the research phase, but then work closely with, and under the leadership of, organizations that represent those most directly harmed by gentrification and displacement.

Disability Justice and Queer Crip Design

So far, this chapter has explored PD as one pathway toward community accountability and control. It turns now to additional lessons from the disability justice movement. As discussed in chapter 1, Disability rights and Disability justice activists popularized the phrase "nothing about us without us" in the 1980s and 1990s.[87] These linked movements have had an extensive impact on the design of everything from the built environment to human-computer interfaces, from international architectural standards to the technical requirements of broadcast media and the internet, and much more. For example, Gerard Goggin and Christopher Newell explore the ways that disability is constructed in new media spaces, as well as how Disabled people have organized to shape those spaces over time.[88] Elizabeth Ellcessor's recent scholarship considers the importance of these movements to the development of media technologies, from closed captioning to the Web Content Accessibility Guidelines, and from the implications of copyright for accessible content transformation to the possibility of collaborative futures designed through coalitional politics.[89]

Over time, disability rights and justice scholars and activists pushed for a shift from the medical model of disability, which locates disability within individual "dysfunctional" bodies, toward the social-relational model: that is, an analysis of how disability is constructed by culture, institutions, and the built environment, which are all organized in ways that privilege some bodies and minds over others. For example, the medical model might seek "solutions" for wheelchair users that would help them stop using wheelchairs, whereas the social-relational model might seek to ensure that buildings, streets, and bathrooms are all constructed to allow mobility for both wheelchair users and non-wheelchair-users.[90] Disability justice work, developed by queer and trans* people of color (QTPOC), has also developed an analysis of the interlocking nature of able-bodied supremacy, racial capitalism, settler colonialism, and other systems of oppression. According to Patty Berne, cofounder and executive director of QTPOC performance collective Sins Invalid, disability justice is built on the principles of intersectionality, leadership of those most impacted, anti-capitalist politic, cross-movement solidarity, recognizing wholeness, sustainability, commitment to cross-disability solidarity, interdependence, collective access, and collective liberation; a disability justice analysis recognizes that "the very understanding of disability experience itself is being shaped by race, gender, class, gender expression, historical moment, relationship to colonization, and more. ... We don't believe human worth is dependent on what and how much a person can produce. We critique a concept of 'labor' as defined by able-bodied supremacy, white supremacy, and gender normativity. ... We value our people as they are, for who they are."[91] Disability justice is to the disability rights movement as environmental justice is to mainstream environmentalism. Scholars, activists, and cultural workers like Patty Berne, the Sins Invalid collective, Alison Kafer, Leah Lakshmi Piepzna-Samarasinha, Aimi Hamraie, and many others have extensively documented this history and have developed tools for intersectional feminist, queer, and Crip analysis and practice.[92]

Another lesson from disability activism is that involving members of the community that is most directly affected by a design process is crucial, both because justice demands it and also because the tacit and experiential knowledge of community members is sure to produce ideas, approaches, and innovations that a nonmember of the community would be extremely unlikely to come up with.

A third key lesson is that it is entirely possible to create formal community accountability and control mechanisms in design processes, and that these can in part be institutionalized. Institutionalization of disability activists' victories proceeded through a combination of grassroots action, lawsuits,[93] policymaking (the Americans with Disabilities Act), and lobbying standards-setting bodies to create and enforce accessibility standards. For these activists, it was important to pressure multiple actors, including lawmakers, government agencies, universities, and private sector firms, to change research and design practices, adopt new approaches, and implement new standards of care.[94] Although these victories are only partial and there is an enormous amount of work to do to deepen the gains that have been secured, disability justice must be a key component of design justice theory and practice.

#MoreThanCode: Findings from the Technology for Social Justice Project

The final section of this chapter explores key findings about community-led technology design practices from #MoreThanCode. *#MoreThanCode* is a PAR report by the Tech for Social Justice Project (T4SJ), meant to amplify the voices of diverse technology practitioners in the United States who speak about their career paths, visions of how technology can be used to support social justice, and experiences of key barriers and supports along the way. The project was coordinated by Research Action Design (RAD.cat) and the Open Technology Institute at New America (newamerica.org/oti), with research partners Upturn, Media Mobilizing Project, Coworker.org, Hack the Hood, May First/People Link, Palante Technology Cooperative, Vulpine Blue, and the Engine Room.[95] From 2016 to 2018, I was part of the coordination team for the project, and I was the lead author of the report that we produced together. As a PAR project, all research partner organizations worked together to develop the research questions, study design, data collection and analysis, conclusions, and recommendations. Over a period of two years, we interviewed 109 practitioners and conducted eleven focus groups, with seventy-nine participants. Interviewees and focus group participants were quite diverse in terms of sexual orientation, gender identity, race/ethnicity, education, geography, and other factors.[96] The first of the report's five key recommendations is "'Nothing About Us Without Us': Adopt Co-Design Methods and Concrete Community Accountability Mechanisms."[97]

Across the ecosystem of social justice and public interest technologists, we found that practitioners know the importance of real engagement with community-based organizations at all stages of the design process. For example, Charley (not their real name), executive director at a technology nonprofit, put it this way: "I think what happens is that people are so quick to say, 'Oh, I got a tool for that.' That's not what we do. We should be listening to the needs of the community. We should be centering the needs of the community over everything else, as our vision. That's sort of like, basic."[98] Study participants from every sector (government, for-profit, nonprofit, and social movement) said that people need to be involved in technology design that is supposed to benefit them. Community-driven design means that communities get to make critical decisions throughout the process; along the way, this approach helps community members develop technical knowledge and skills. For example, Heiner (executive director of a legal service organization) emphasized the importance of having people who are poor, undocumented, seeking housing, and/or have dealt with the criminal justice system at the table when creating civic tech apps that are supposed to be for them. Hibiki, a digital security trainer, put it this way: "[Community-led design is] all about developing tools and technology along with the people that it's meant to serve. Just, in general, I think adopting any type of participatory approach from the beginning is usually super helpful, and also enables people to actually want to use this technology."[99] One concrete accountability mechanism that practitioners suggest is community advisory boards or governing councils that can guide and own design processes.

In contrast, civic tech projects that lack community leadership tend to fail. Several interviewees used civic gamification platforms as examples. Hardy, a technology capacity builder and crisis response specialist, said: "[These] platforms tried to get people engaged with civic planning without understanding that they had to be able to implement what people were talking about. You can't just ask people for their opinion. You also have to act on their opinion."[100] Even when there is a clear need for a new tech solution, community-specific user research should precede design and development. According to Lulu, a funder at a national foundation: "We funded an earned income tax credit tool [because] ... unfortunately billions of dollars each year go unclaimed by the working poor because they don't know they're entitled to it. So, we built a system like that, and it got a lot of usage in English, but when we built it in Spanish and Vietnamese almost nobody used it. ... So either we don't understand how to deliver technology to these special language groups, or we're not doing the right outreach, or it's not culturally appropriate, I don't know."[101]

Projects with good intentions are not immune from failure, and can even cause inadvertent harm. Alda, a program manager at a national organization, helped build an SMS voter registration system, but the project team then built a voting component into the tool that had the potential to expose community members' voting history: "It was kind of just built because it could be built. ... There was no analysis on the political context of what could happen if they started using that and different groups got hold of telecoms and could ask telecoms to turn over that data. SMS is clear text. It's very easy to see then who you voted for, depending on what your mobile number was. There's just so many things wrong with that. I feel like that was something built with good intentions, but they did not do any of the risk modeling that they should have done."[102] Thinking that tools provide silver bullets without taking the time to really understand community contexts is also a recipe for failure. #MoreThanCode study participants shared many stories of failed projects; frequently, those projects jumped too quickly to "solutions" and were tool-centric. At best, this approach tends to waste scarce resources and time. Tivoli describes one failed project that stood out for her as a user researcher—an iPad-based self-assessment tool for elderly people: "It completely failed, because it was a technology solution. I don't remember if it was the same group that redid it or if it was a parallel project, but someone did a brochure, and it was much more successful. ... We don't have to always make an app for it."[103]

Before deploying new tools, it's essential to verify whether they meet organizational needs. Although organizations may be eager to adopt new technologies, pushing the wrong tool can result in backlash, mistrust, and, over the long run, even greater inefficiency. For example, one practitioner described an example in which a consultant foisted an unnecessarily complicated new database on an organization, and staff became so frustrated that they abandoned databases altogether and went back to time-consuming paper processes. What ultimately matters is not tool adoption: it is people's struggles and their lived experiences. Gertruda, a digital security researcher, put it this way: "The struggle is not access to encryption tools. It is organizing day labor communities in order to protect against ICE raids and things like that. We're confusing means and ends."[104]

We also heard from many practitioners that funders tend to support "techie parachuters" for a quick fix, instead of investing to build capacity within a community. These quick fixes are not sustainable, as Charley notes: "We have funders that will fund large organizations who have large amounts of money to fly in to communities of color and basically tell them, this is how things should be done. We disagree. I disagree with that methodology and that strategy. One is that there are people within the communities already with knowledge, or lots of knowledge, who are not being lifted up. Two, we believe that if we're really going to build power, we need to build power in the communities, which means we need to let go of our ego and we need to sort of build, mentor, build that power in the community, build the skills there."[105] Several practitioners said that funders need to listen to community organizers, not only to techies. However, too frequently designers and technologists who occupy privileged positions within the matrix of domination influence funder decisions about who should receive resources. They also get to decide what tools are considered "cool," without much consideration for community context or the broader implications of their preferred approach.[106]

After talking with designers, developers, researchers, community organizers, funders, and other practitioners around the country, the T4SJ Project synthesized hundreds of concrete suggestions for community accountability into the following recommendations:

Adopt codesign methods. This means spending time with a community partner, in their space, learning about needs, and working together through all stages of design. Usually, no new tech development is necessary to address the most pressing issues. Codesign methods have a growing practitioner base, but they could be better documented.

Develop specific, concrete mechanisms for community accountability. Nearly all interviewees said that the people most affected by an issue have to be involved throughout all stages of any tech project meant to address that issue. All actors in this field need to move past stating this as a goal and toward implementing specific, concrete accountability mechanisms. For example: funders should require concrete community accountability mechanisms from their grantees, and educators should center community accountability in education programs.

Center community needs over tools. Community needs and priorities must drive technology design and development, and technology is most useful when priorities are set by those who are not technologists. Be humble and respect community knowledge. Process and solution should be driven by the community; do not make community members token participants.

Invest in education (both formal and informal) that teaches codesign methods to more practitioners. Support existing efforts in this space, create new ones, and push existing educational programs and institutions to adopt codesign perspectives and practices.

Create tech clinics, modeled on legal clinics. Public interest law and legal services work are client-oriented, and lawyers doing this work are constantly interacting with people who need to navigate larger unequal systems. This is considered part of their legal education. Tech can learn from this model.

Avoid "parachuting" technologists into communities. In general, parachuting is a failed model. Don't do it. Stop parachuting technologists into organizations or focusing on isolated "social good" technology projects, devoid of context, when the real need is capacity building. We are not saying "never bring someone in from outside a community." ... We do think that it is worthwhile to develop better models for sharing local knowledge with national groups and for national groups to share their perspectives with local groups in such a way that all parties can benefit.

Stop reinventing the wheel! Well-meaning technologists often reinvent the wheel, without researching existing solutions. Designers, developers, and project leads, no matter what sector they are in, should begin projects by researching existing projects and organizations. This also stems from competitive, rather than collaborative, mindsets ("ours will be better, so we'll just compete"). It is important to work together to develop shared tools and platforms, instead of only competing for scarce technology resources.

Support maintenance, not just "innovation." Significant resources are necessary to maintain and improve existing movement tech, but most focus is on the creation of new projects. We need more resources to update, improve, and maintain already proven tools.[107]

Conclusions

Ultimately, although all people design, only some people are employed as design professionals. Unfortunately, access to paid design work is deeply unequal and is shaped by the matrix of domination. Although the larger problem is structural, individual design firms can help if they develop inclusive hiring and retention plans, publicize specific targets and dates for staff, leadership, and board diversity, adopt best practices in accountable community partnerships, and share profits and credit with community partners.

Beyond employment equity, design justice requires full inclusion of, accountability to, and ultimately control by people with direct lived experience of the conditions the design team is trying to change. Not only is community leadership ethical, but also, the tacit and experiential knowledge of community members is sure to produce ideas, approaches and innovations that no one else would be able to create. People's lived experiences of race, class, gender identity, sexual orientation, disability, immigration status, language, age, and so on structure variance in user product needs, as well as access to the resources that are needed to address those needs. There are several approaches to design practice that recognize these dynamics and attempt to address them, at least in part. These include human-centered design, participatory design, and codesign, among others.

Human-centered design (HCD) includes end-users in the design process through various strategies; it focuses on better-matched affordances and improved user experience. This is good—but it has little to say about values, community accountability or control, or the ultimate distribution of benefits such as profits or attention. It may be used by any institutional actor, and it may be used for extractive design processes that gather ideas from marginalized communities, create products, and sell them back to that community (or elsewhere). Participatory design (PD) and codesign, in contrast, attempt to include end users throughout the design process. Most PD processes also aim to develop feelings of investment and ownership in the outcome by all participants, and many PD practitioners are also deeply concerned with questions of community accountability. However, the discourse of PD has in some cases been co-opted, on the one hand by university-based researchers, and on the other by multinational firms, governments, and other powerful institutions. PD, like HCD, is sometimes used for extractive processes that gather community input but primarily produce benefits for the careers of professional design researchers and practitioners. PD processes sometimes, but not always, have formal community accountability mechanisms, and do not always center community power and control.

Design justice is aligned with and draws from the history of PD, and may employ specific techniques from PD, codesign, and HCD. However, it also focuses on concrete mechanisms for community control, is linked to a disability justice analysis, and explicitly attends to the distribution of design's benefits and burdens according to the matrix of domination. Design justice proposes a shift in the unaccountable and deeply inequitable state of affairs in design practice at several levels, including toward a more inclusive professional design workforce, as well as recognition of and resources for community-led, Indigenous, and diasporic design practices. This requires work at many levels, from micro to macro, from individual design projects all the way up to transnational standards bodies.

At the micro level, individual professional designers from various design fields can learn how to participate in community-led processes by bringing our skills and resources to the table, rather than seeking community members to participate in processes that we initiate and control. Design teams can adopt strategies for community accountability and control, such as inclusion of community members with direct lived experience of the design problem, intersectional user story validation and testing, and formal memoranda of understanding (MOUs) or working agreements that set clear expectations about project roles, decision making, and ownership of design products. Formal agreements are especially important when working with historically marginalized communities but also apply to any design process. A design justice framework also requires gathering resources to enable meaningful community participation and shared ownership.

Design justice can help guide us in the long-term struggle to transform institutions such as professional associations, universities, and standards bodies so that they are more accountable to communities that are marginalized within the matrix of domination. In universities, a design justice approach can shift the way design is taught (chapter 5) and help develop a generation of designers who practice community leadership, accountability, and control. Standards bodies can adopt and promote standards that include community accountability, as well as an intersectional approach to benchmarks, testing, and audits. At the level of the nation-state (as long as nation-states exist), we need policy changes to shift priorities toward research and design that center the needs of historically marginalized communities, incentivize formal community accountability and control mechanisms, discourage extractive approaches to design work, and provide far greater resources for already-existing networks of community-based design practitioners.

3 Design Narratives: From TXTMob to Twitter

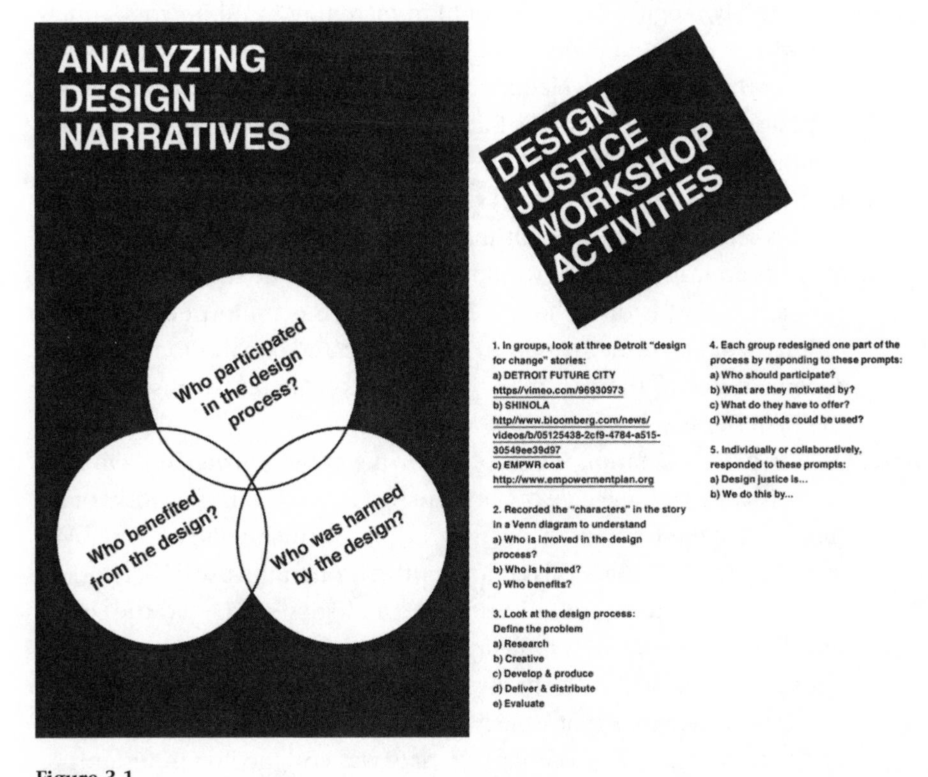

Figure 3.1

"Analyzing Design Narratives," from *Design Justice Zine*, no. 1: Principles for Design Justice (ed. Una Lee, Nontsikelelo Mutiti, Carlos Garcia, and Wes Taylor). Designed by Nontsikelelo Mutiti and Alexander Chamorro. Available at http://designjustice network.org/zine.

> Designing is not a solitary activity. It is a part of a larger social community of discourse.
>
> —Drew Margolin

> People are aware that they cannot continue in the same old way, but are immobilized because they cannot imagine an alternative. We need a vision that recognizes that we are at one of the great turning points in human history when the survival of our planet and the restoration of our humanity require a great sea change in our ecological, economic, political, and spiritual values.
>
> —Grace Lee Boggs

It is Sunday, August 29, 2004, and I'm marching in midtown Manhattan with a crowd of more than half a million people during protests outside the Republican National Convention. Most are there to voice opposition to the US war in Iraq, at a rally organized by the antiwar coalition United for Peace and Justice. The invasion of Iraq and Afghanistan, and the so-called war on terror, launched by George W. Bush in 2003 based on what would ultimately be shown to be false information about chemical weapons,[1] would drag on to become the longest armed conflict in US history. The war cost at least 5.6 trillion dollars,[2] with estimates of between one hundred thousand and one million casualties to violent death, the vast majority of them civilian.[3] At this moment, the Iraq War is still less than two years old.

Today's mobilization is part of a growing cycle of struggles.[4] On February 15, 2003, global civil society and social movement networks organized the largest simultaneous day of protest in human history.[5] We were able to coordinate this action partly through the use of networked information and communication technologies (ICTs; at the time, mostly email lists, Internet Relay Chat, and Indymedia open publishing sites), combined with the strong personal and organizational social movement networks that we developed over two decades in the global justice movement. The February 15 date was cosigned by thousands of organizations from around the world, during the World Social Forum that took place the previous January in Porto Alegre, Brazil. Although we failed to stop the war before it began, in the early days of 2004 it still seemed possible to many of us that the US presidential election might be an opportunity to quickly end the war.

The massive march has been entirely peaceful, but there is a tense atmosphere because our path is controlled by hundreds of police officers at multiple intersections. I am marching with my sister and parents. We reach a spot where the crowd has come to a standstill. Police with batons block our way. Suddenly, a surge of bodies pushes us backward as a line of officers mounted on horseback rides single-file through the crowd; everyone scrambles to get out of the way. A few feet from us, one of the mounted police suddenly rears his horse up onto its hind legs; hooves fly through the air, dangerously close to an elderly woman who cries out and ducks for cover. My sister, Larissa, grabs onto my arm, and one of us says something like "What the fuck?!" We back away quickly and try to make our way to a less chaotic part of the mobilization.

Police have already arrested hundreds of protesters during the previous two days; on Thursday, twelve ACT UP! activists were arrested for a naked protest against Bush's regressive global AIDS policies. On Friday, 264 people were arrested during a huge Critical Mass bicycle ride of five to six thousand riders.[6] Overall, during the course of the convention and the protests, more than 1,800 people, including protesters, bystanders, legal observers, and journalists, will be arrested, fingerprinted, and held in makeshift pens in a toxic former bus depot. The vast majority (more than 90 percent) will face charges that will be dropped or thrown out of court, and ultimately (ten years later, in 2014), New York City will settle a class-action suit by the ACLU for nearly $18 million—the largest protest settlement in US history.[7]

After another hour or so, I say goodbye to my family and make my way to the makeshift, semiclandestine Independent Media Center (IMC, or Indymedia) that has been set up to cover the protest. The IMC is a hub of frantic activity. In one corner, a young woman imports footage of police violence from at least three different kinds of handheld video cameras (mini DV tape, hard disk drive, and VHS-C) into the editing software Final Cut Pro. Some of this footage will be uploaded quickly to Indymedia (YouTube does not yet exist); some will be used later by legal support teams to ensure that most of the arrests are thrown out of court (and still later as evidence in the class action suit); some will be used to produce documentary films about the event, such as *We Are Many*.[8] In a side room, a small team works to produce audio for a podcast and to send clips to various radio stations affiliated with the listener-supported Pacifica radio network. My task is to gather and confirm reports of various actions, arrests, and incidents of police brutality that are coming in from across the city via phone calls, emails, text messages, and uploads to the Indymedia open publishing newswire.

As I do this, to remain in close coordination with other media activists around the city and around the world, I'm logged in to the Indymedia Internet Relay Chat (IRC) server and participating in several relevant channels. IRC channels are dedicated, persistent, chat-based conversations, marked by the pound sign—for example, #RNCarrests for conversations about arrests at the Republican National Convention. The # (pound sign or hash) marker for conversations on activist chat servers would later make its way into much broader use in the now-ubiquitous social media feature we know as *hashtags*.[9] It should not be surprising that the ability to create ad hoc groups, or ongoing conversations, instantly with the pound sign was pioneered by hackers and activists, and yet today this is not widely known. On IRC, I receive a message from a friend who uses my handle, @schock, to notify me; using the @ (at) sign to notify a particular user in a channel that you have sent them a message is another feature that was imported from IRC into many social media platforms today. He wants to know whether I have successfully signed up for TXTMob.

TXTMob is an experimental group short message service (SMS) application that was developed by design professor Tad Hirsch, who at the time was a graduate student at the MIT Media Lab.[10] At the RNC in New York, hundreds of people, most of them seasoned activists, used TXTmob to coordinate, share verified information about actions in the streets, and keep abreast of police activity. Although it was designed to work via SMS and therefore could be used on nearly any mobile phone (remember that almost no one had a smartphone in 2004), it was not widely adopted beyond activist circles. It was a student project, with poorly written code, and it used a clunky hack to send SMS for free: it took advantage of the email-to-SMS gateways that nearly all mobile operators made available at the time. Indeed, if hundreds of thousands of protesters had all signed up for TXTMob, the tool quickly would have been blocked by mobile service providers once they noticed the volume of messages being sent without payment. In any case, TXTmob mostly worked. It provided a useful information sharing service to its small group of highly connected activist users. It helped increase the circulation speed of verified information, helped direct action affinity groups make tactical decisions about which street corners to blockade, and helped confirm key developments and dispel some of the false rumors that tend to spread like wildfire during mass protests.[11]

After the RNC was over, Tad Hirsch met with Gaba Rodriguez, Rabble, Blaine Cook, and other activist developers at the Ruckus Society SMS Summit in Oakland to talk about the state of SMS tools for activism, including what had worked well at the RNC and what needed improvement.[12] For their day jobs, Gaba, Rabble, and Blaine worked at Odeo, a podcasting startup that was rapidly running out of seed money. Although the company had a decent product, there were not enough people creating or listening to podcasts at the time to create a sustainable business model. The death blow came when Apple announced that iTunes would soon launch a podcasting product. With only enough seed money to pay for a few more months of payroll, the Odeo employees decided to mostly abandon work on their main product and switch over to hacking on other potentially interesting projects that might be able to attract new investors or spin off into their own companies. To kick off this process, Odeo held a demo day during which various teams put together project ideas, presented them, and then decided what to work on for their remaining salaried time.

One team, led by the hacker-activists who had been part of the RNC protests, presented TXTmob. They talked about the tool in the context of the protests: what had worked well, what failed, and what features of the tool might be compelling for a broader set of possible users. For example, account creation and group signup were both very clunky in TXTmob, so those would have to be improved. The method of sending SMS via telecommunications company (telco) gateways wouldn't scale beyond a few hundred or a few thousand people, so that would have to change as well. However, the team argued, there was a lot of potential in a group SMS application focused on providing real-time updates. Others at Odeo agreed. Over the next few weeks, TWTTR (Twitter's original name) was born, and as they say, "the rest is history."[13] In the context of design justice, however, we must ask: *Whose version* of history?

The story that I have just narrated about the origins of Twitter is not widely known. Instead, as Hirsch writes: "Nick Bilton's October 13 New York Times Magazine story, 'All's Fair in Love and Twitter,' describes the heady, early days of Twitter. The article begins with [Twitter cofounder] Jack Dorsey sitting atop a slide in a 'rinky dink' Silicon Valley playground sometime in 2006, expounding his vision of a microblogging platform to a handful of Silicon Valley techies and entrepreneurs who would go on to create one of the most popular web services in the world. ... It's a compelling story. Unfortunately, it isn't true."[14] Hirsch, who is now the chair of Art and Design at Northeastern University, is not interested in claiming that he is the "actual" inventor of Twitter. Instead, in a clear and compelling article that is worth quoting at length, he describes his interest in setting the record straight:

> To be clear, TXTmob wasn't Twitter. The Twitter team made a number of key innovations that allowed the project to scale, and to attract investors. Further, pointing out that TXTmob played a role in Twitter's creation is in no way to suggest that Evan, Blaine, Jack Dorsey, or anyone else stole anything from me. TXTmob was an open-source project that I freely shared. The folks at Odeo took this project and adapted it for mainstream use in ways that I frankly did not anticipate. And while I wouldn't object if one of the Twitter millionaires decided to send along a few "thank you" shares, I don't believe that they are under any obligation to do so ... However, I do think it is important to get the story right. As Bilton observes, creation myths matter. They don't simply tell how things happened, they tell us who we are. Jack Dorsey clearly needs to believe that he's not just clever (and lucky), but that he's a rare breed of genius. It's also probably important to Twitter's employees and investors to believe this too. The problem with Dorsey's story, for the rest of us, is that it describes a world where the market is the sole site of technical and social innovation, and where we are wholly dependent on a handful of extraordinarily gifted entrepreneurs to lead us out of the dark ages. This is a myth. The truth is that Twitter—or something very nearly like it—would almost certainly have happened without Jack Dorsey. However, it might very well not have happened without the long progression of earlier tinkerers and dreamers, who often worked well outside the confines of the market. Their collective efforts paved the way for many of the technical marvels we now enjoy, and we should take care to ensure that they are not written out of the histories of the extraordinary age in which we are living.[15]

This chapter is about how design narratives provide an important arena of contestation for the theory and practice of design justice.

Design justice means that we consider the values that we encode in the objects and systems we design, as we discussed in chapter 1, as well as who gets to participate in and control design processes, as we discussed in chapter 2. It also means that we think about design narratives: who receives attention and credit for design work, how we frame design problems and challenges, how we scope design solutions, and what stories we tell about how design processes operate.

Smart Men and Start-Ups: Innovation, Attribution, and Appropriation

What is innovation, beyond a buzzword? There is a burgeoning corporate literature that promises to reveal the "secrets" of innovation, full of titles like *The Innovator's DNA: Mastering the Five Skills of Disruptive Innovators*[16] and *The Art of Innovation: Lessons in Creativity from IDEO, America's Leading Design Firm.*[17] There is also considerable attention to the subject within the academy. Subfields of economics, management studies, and design, as well as urban studies and planning, anthropology, sociology, and science and technology studies seek to better understand various aspects of innovation and innovators. Works in the history of science and technology often unpack how a particular technological innovation unfolded over time.[18]

Popular narratives about innovation are dominated by the figure of the genius. In popular culture, we are often led to believe that all technology is created by brilliant, well-educated, mostly white (cis)men, working in university labs, corporate R&D departments, or perhaps in their garages, who go on to found Silicon Valley start-ups. This narrative is tightly entwined with the mythology of meritocracy: people get what they deserve, and if you work hard enough, you will achieve your dreams.[19] Yet, as sociologist Robert Merton argued in 1968: "There is often a disjuncture between America's meritocratic values that promote aspiration for success and the opportunity structure—the social, economic, and political structures that make success possible. The problem is that opportunities are not equally distributed, and they are not allotted solely by meritocratic criteria. For example, racism serves as a strong barrier to African American's achievement. Even if unintended, the promise of equality inherent in meritocratic ideology serves to elide racism."[20] The opportunity structure is not only raced, it is also gendered, as feminist legal scholar Deborah L. Rhode has described.[21] Even as access to key jobs in the information economy is structured by linked white supremacy, heteropatriarchy, class inequality, ableism, and other aspects of the matrix of domination (as discussed in chapter 2), this reality is obfuscated by the mythology of meritocracy. In other words, many white (cis)male technologists believe that their position as "innovators" and access to the attendant benefits (salaries, titles, credibility, prestige) are based primarily on their raw talent and individual brilliance. However, access to these positions is shaped by structural inequality, even as sociotechnical innovations frequently emerge from marginalized communities but are then appropriated by powerful actors. Indeed, user innovation is the norm, not the exception to the rule.

Still, the *diffusion of innovation* remains the most widespread theory about how innovation works.[22] In this model, innovators (scientists, researchers, inventors, technologists) create a "new technology." Over time, if it is a useful invention, this new technology "diffuses" or spreads out from the epicenter of its site of invention. It is taken up first by "early adopters," then moves into broader distribution, and finally is adopted by nearly everyone, save a few holdouts or laggards. The model is illustrated in figure 3.2. Although this model remains influential, scholars of science and technology have challenged it on several grounds. First, it contains a somewhat masked normative assumption that "technology adoption" is always a good thing. To illustrate, simply imagine this model applied to a technology that is widely recognized to be harmful—for example, personal ownership of military assault rifles, or crack cocaine and crack pipes. Second, it has nothing to say about the many factors that might influence the adoption of desired technologies—most obviously, wealth disparity, but also gendered and raced cultural norms, among many other variables. Third, diffusion theory imagines technologies as static, although technological objects and the ways they are used (sociotechnical practices) are constantly changing. Early versions of a new technology are nearly always quite different from later mass-market versions. Not only is innovation iterative, but many, if not most, small changes (iterations) to a given technology are made by everyday people (users), rather than by professional scientists, researchers, or product designers.

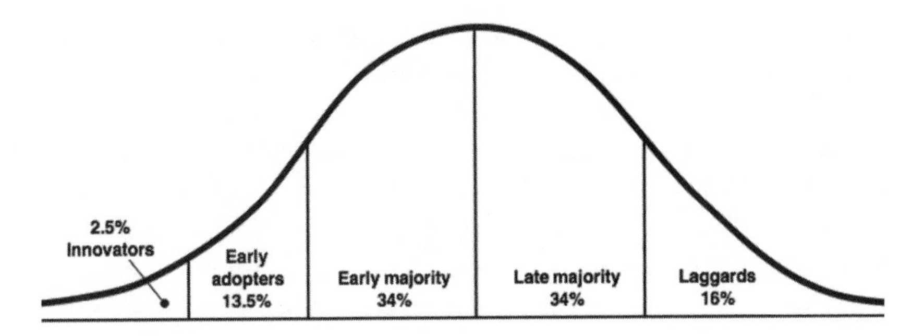

Figure 3.2
Diffusion of innovation. *Source:* Rogers 1962.

The prevalence of user modification is the core insight of the theory of technology appropriation. As technology scholars François Bar, Matthew S. Weber, and Francis Pisani put it, "Appropriation is the process through which technology users go beyond mere adoption to make technology their own and to embed it within their social, economic, and political practices."[23] For example, they trace the history of mobile money, which began as a user innovation, to illustrate what they call the *cycle of technology appropriation*. Initially, Kenyan mobile users appropriated prepaid top-up cards and repurposed them as a form of payment: they purchased cards, then sent the card numbers to other users via SMS. Later, mobile phone companies noticed the user innovation and launched mobile payment services (like M-Pesa) bundled with their phones. The authors argue that user appropriation is a key, but often overlooked, part of most innovation processes.[24]

The theory of technology appropriation is similar to lead user innovation theory, discussed in the previous chapter. In his text *Democratizing Innovation*, MIT management professor Eric Von Hippel both theoretically and empirically demonstrates that a significant portion of innovation is actually done by users, rather than manufacturers.[25] Further, he finds that particular kinds of users (lead users) are the most likely to innovate and that their innovations are more likely to be attractive to a broader set of users. In design justice terms, those whose needs have long been marginalized within the matrix of domination have a strong information advantage when it comes to articulating those needs and developing possible solutions. In terms of information costs, a user, user community, or small organization rooted in a marginalized community thus is often best positioned for innovation. This is due both to the high amount of specialized domain knowledge they possess and to the low costs of testing possible solutions in the real-world "laboratory" of daily life.[26] This means that knowledge "extraction" is not only unjust, it is also costly and inefficient. Von Hippel makes a compelling generalized case for why manufacturers and users tend to innovate based on the information that they already possess and for why lead users should be included in design processes. Design justice extends this observation to consider the ways that the matrix of domination (race, class, gender identity, sexual orientation, disability, age, language, and so on) structure participation in and exclusion from product design, development, and manufacturing. In Von Hippel's terms, the difficulty of sharing both use context and solution information decreases the likelihood of product innovation that centers the specific needs of marginalized users; information stickiness suggests that users with lived experience of the design challenge should be incorporated into any design team.

Attribution: Giving Credit Where Credit Is Due

Von Hippel provides a compelling argument with great potential to shift larger narratives about technology design. However, in practice, his work has mostly been used to help firms develop strategies to encourage lead user innovation, then appropriate user innovations to increase their own profitability. In contrast, one of the key principles of design justice is full attribution. Under white supremacist capitalist heteropatriarchy and settler colonialism, the work, ideas, inventions, belongings, land, and very bodies of women, Black people, people of color, and indigenous peoples have been appropriated (stolen) for centuries by wealthy white (cis)men. This process is perhaps most extreme, most visible, and most acceptable to talk about today as something that took place historically during the age of colonialism, empire building, and the transatlantic slave trade.

However, the theft of others' labor, time, energy, culture, innovations, and ideas, as well as land and bodies, by those in positions of structural power continues today. Design practices, norms, and institutions are by no means immune to this dynamic. Instead, as design becomes increasingly central to economic, cultural, and social life, if anything there is increased pressure toward appropriative design strategies. As design gains power, the stakes become higher. Design jobs are increasingly lucrative, and competition for contracts, investors, "intellectual property,"[27] and visibility are fierce.

The typical capitalist firm is arranged in a pyramid structure, so that resources (time, energy, credit, money) flow from bottom to top. This is also the case within most design firms. At the extreme, in large multinational design enterprises, armies of poorly paid underlings labor to produce work (concepts, sketches, prototypes), while the benefits (money, attribution, copyrights and patents) flow upward into the hands of a small number of high-profile professional designers at the top. There's also a power law at play, as in every industry. A few highly visible design firms and individual designers reap massive rewards, while the expanding legions of unknown firms and designers struggle to make ends meet. Of course, the "winners" in the power law game in design, as in every industry, aren't random, and it's not a true meritocracy. Instead, the design field is gendered, raced, classed, and otherwise shaped by (and shapes) broader conditions of structural inequality. The scenario that we find dramatized in *Mad Men*, in which women workers in an advertising firm are occasionally called into the office so that they can be briefly "mined" for ideas, or in which even when promoted to designer, their ideas are passed over or stolen by men, endlessly repeats.[28] In addition, the internet has enabled a new scale of extractive mechanisms in design. For example, this is often this case in "design challenges" in which dozens, sometimes hundreds, of people do free labor and submit ideas in hopes that they'll be the lucky one chosen to receive visibility, recognition, and possibly even compensation. Most recently, this process has been platformized, in spaces such as OpenIDEO, DiscoverDesign, and IdeaScale.

At its most basic, the principle of attribution simply says that design justice includes giving credit where credit is due. This principle applies across the life cycle of the design project, includes any products, and should also shape the story of the project as it is told to various audiences. In design justice, those whose lived experience guides the process are recognized as codesigners; they become co-owners of designed products, platforms, systems, and other outputs and also become coauthors of the story about the project.

Design justice considers "Who contributed?" to be a critical question for the evaluation of any given design project. Although it may not be necessary to invent new mechanisms for attribution to give credit where credit is due, there are some recent experiments in this direction. For example, J. Nathan Matias's project thanks.fm was an attempt to make it simpler for people to share credit with others on the web.[29] Although it served largely as a speculative design project and is no longer active, it helped call attention to full attribution as an (intersectional feminist) design principle. However, ultimately, attribution is not primarily a technical problem: it is a social and cultural one. In a similar vein, Anil K. Gupta's Honey Bee Network recognizes local inventors by name.[30] Gupta founded the network in part out of his frustration at the extractive knowledge processes of many so-called development projects. Black feminist cyberscholar Kishonna Gray created the #CiteHerWork hashtag to address the theft and erasure of (especially) B/I/PoC women and femmes in academic and journalistic writing communities, including those that analyze technologies and sociotechnical practices.[31] Science and technology scholar André Brock developed *critical technocultural discourse analysis*, a mode of analysis that uses critical race, feminist, and queer theory to unpack how marginalized users often produce technocultural practices that become the core use case for digital tools and platforms, with Black Twitter as a key case study.[32] Much more work remains to be done to mainstream these and other approaches to proper attribution in design.

Equitable Distribution of Attention in Design Processes

Design justice is concerned with the equitable allocation of benefits and harms that result from any design project, and the design of new technologies often produces discursive benefits and harms. In other words, the stories that we tell about design matter. As media scholars Sarah Jackson, Moya Bailey, and Brooke Foucalt-Welles note, "Discourse constructs reality by making ideas and events meaningful in particular ways that uphold, and/or challenge cultural ideologies."[33] Scholars such as Thomas Davenport and John Beck argue that we now live in an *attention economy*,[34] where mediated visibility has become an important form of capital. Attention (time) is a scarce resource within late-stage informational capitalism, and its allocation has significant symbolic and material impacts. Design projects command attention: both social and mass media are full of stories about the latest designed objects and about the people and firms that design them. Certain individuals, organizations, and communities gain attention as designers, innovators, technological creators, or (especially in stories about technology and marginalized communities) "saviors." However, for a variety of reasons, marginalized individuals, communities, and movements rarely receive much of this attention.

Why is this the case? First, as discussed in chapter 2, paid professional design work as an elite job field is deeply unequal along race, class, and gender lines. There are fewer LGBTQI+ and/or B/I/PoC designers overall, fewer still who occupy powerful positions within design industries, and therefore fewer stories about marginalized people's design accomplishments. Additional dynamics are at play that even further distort design discourse: for example, elite networks of technology reporters and tech industry insiders; class, race, and gender dynamics within the journalism profession, especially in tech journalism; and so on. On the other hand, there is a kind of cottage industry, or at least minor narrative, that focuses on "surprising" examples of technological design and innovation by marginalized people.[35] Without entirely dismissing this genre, it is typically quite tokenizing; these stories also often reinforce normative gender narratives about women's roles. Such stories very rarely engage in deeper conversations about *why* it should be surprising to see, say, a start-up led by an all-woman/Black/Indigenous/queer/trans* (and so on) team.

Put simply, design projects generate attention, attention is valuable, and design justice as a framework thus asks us to explore whether this attention is equitably distributed. How can design teams ensure more equitable attention distribution? Concretely, there are many possible strategies: design teams can include clauses about attribution in MOUs; take care to name community partners in press releases, reports, and all materials that describe the project; provide attribution to community partners in patents, licenses, and software-release notes; and consider how to allocate opportunities to speak about the project to journalists and other potentially interested audiences, such as policymakers, academics, and funders. Also, new kinds of attention analytics can be used to evaluate design projects within a design justice framework. Rather than focus on the raw number of news stories, quotes in prominent outlets, or social media metrics such as shares, likes, and comments, design justice practitioners might analyze stories about their projects to better understand how attention has been allocated, whose voices are heard most frequently, and whether that allocation fits the design team's goals and principles.

Resistance Is Fertile: Social Movements, Media Innovation, and Corporate Appropriation

If most design is lead user innovation within particular communities of practice, why do cultural narratives about individual genius inventors predominate? In part it's because the most visible narratives about design and innovation are well-resourced corporate mythologies. An entire industry of technology "reporting" has been built largely on press releases from established firms, start-ups, and venture capital–backed incubators. However, there are many other ways to narrate the history of technology design. One approach is to focus on the contributions of social movements.

Social movements have long been a hotbed of innovation in media tools and practices, in part because of their relationship to the media industries. As the slogan says, "Resistance is fertile!" Movements, especially when led by marginalized communities, are systematically ignored and misrepresented in the mass media, so movement organizations often develop strong media practices, active counterpublics, and innovations in media technology out of necessity.[36] Many social movement media innovations are later adopted by the journalism profession and by the cultural industries. Indeed, this happens so frequently that political scientists William Gamson and Gadi Wolfsfeld theorize social movements and the media as interacting systems.[37]

This chapter began with the story of TXTMob and Twitter, but many similar tales could be told. Social movement design innovations include media technologies, but also decision-making processes, tactics for pressuring elites, and policy proposals, as well as cultural, artistic, and aesthetic forms. For example, the diverse, interlinked social movements in the United States in the 1960s and 1970s, including Black and brown power, feminist, antiwar, Disability, and other movements, produced and influenced a wide range of cultural innovation in fields like music, painting, film, dance, and more. Media historian Fred Turner has argued that 1960s movement counterculture led to broader social transformations and to the development of the internet.[38] Many movement-led innovations from this time period provided fodder for the reinvigoration of mainstream cultural industries. In the late 1990s and early 2000s, media innovations by the global justice movement played a similar role in seeding ideas, demo designs, and proofs of concept for participatory media making that would later become the core of "culture industries 2.0." Soon, as media scholar Tiziana Terranova put it, nearly everyone would be performing "free labor for the digital economy."[39]

The Misrepresentation of Social Movements in the Mass Media

In part, social movements are important spaces for media innovation because activists are so badly misrepresented by the mass media system. Empirical studies of mass media coverage of social movements bears out activists' lived experience: print and TV news provide little sustained coverage of social movements, and when they do, rarely adopt movement framing.[40] In particular, mass media tend to cover protests using violent conflict frames.[41] This downplays the arguments and legitimate grievances of protesters in favor of spectacular imagery and language about violent conflict between protesters and police. Even as protest policing has become increasingly militarized,[42] mass media have begun to embed reporters with police; this results in coverage that is systematically biased toward the perspectives of law enforcement. In the wake of the failure of US military and information policy in Vietnam, the US government developed more sophisticated practices of information shaping and control to avoid losing "the media war" during future imperial adventures. Many of these tactics were deployed and honed during Gulf War I, in which information control played such an important role that philosopher and media theorist Jean Baudrillard famously described the war as a *simulation*.[43] The practice of "embedding" reporters with US military units in Iraq and Afghanistan, widely discussed in the 2003 invasion of Iraq,[44] was subsequently deployed during domestic protest policing in the United States.[45] For example, since 2003, mainstream media outlets have embedded reporters with police units during most large-scale protests, including the 2003 mobilizations against the Free Trade Area of the Americas in Miami; the 2008 protests at the Republican National Convention in St. Paul/Minneapolis; the

Democratic National Convention in Denver; the 2014 protests in Ferguson, Missouri;[46] and many other large-scale protests since.

The Revolution Will Be Livestreamed

Under conditions of scarce and poorly framed coverage from powerful media organizations, social movements have always taken it upon themselves to self-represent. Indeed, the history of the early print press is in large part a history of social movements, political parties, and ethnic groups each producing their own newspapers.[47] To take a more recent example, the last decade has seen widespread adoption of livestreaming. From early on, livestreaming tools were appropriated by social movement actors, who also often innovated new approaches to the way the technology was used. Activists freely shared these innovations through social movement networks, and some of these innovations were then incorporated into new iterations of livestreaming platforms, products, and tools. Examples of livestreaming by social movements include the antinuclear movement, livestreamed by Deep Dish TV (via satellite, in the 1980s);[48] the Global Justice Movement, livestreamed by Indymedia during the early 2000s (via DIY servers); the immigrant rights movement, which used Ustream and livestream.com to transmit sit-ins from five congressional offices in July 2010 (the first time this had ever been done);[49] Occupy Wall Street (streamed by Global Rev and others);[50] the Brazilian antiausterity mobilizations, streamed by Midia Ninja;[51] and the many livestreamers within #BlackLivesMatter and other more recent movements.[52] In the earlier examples, activists organized their own livestreaming infrastructure with handheld cameras, Linux laptops, and free/libre software (usually ICEcast and/or VideoLAN) and maintained their own streaming servers. Later, most switched to commercial livestreaming video services such as Bambuser, then Meerkat and Periscope, then Twitter and Facebook Live, then Instagram Live and Twitch.[53]

Radical Technology Collectives: Autistici/Inventati, RiseUp, May First/People Link, and Beyond

Although much social movement ICT innovation happens "organically" around the edges, as activists cobble together whatever they need, there is also a long history of dedicated tech activists who organize radical tech collectives (RTCs) to support movement organizations more systematically. Recently, the Italian RTC Autistici/Inventati (A/I) published an English translation of its history, composed of interviews with collective members and contextual notes. This book provides a detailed and fascinating history of Italian hacker and media activist projects and collectives, the ways that they were linked to social movement organizations, their constant evolution, fragmentation, recombination, and adaption to the changes in networked communication technology from BBS systems to the web, their integration of GNU/Linux and free software, the role that they played in developing and promoting encryption among social movement groups, and much more.

In the preface, media activist and theorist Maxigas describes the dynamics of politicized hacker culture. In Europe, he notes that there are three overlapping but distinct circuits, or scenes, of political hackers; one in Northern and Western Europe, more focused on technical innovation; another in Eastern Europe and the Baltic states, centered on the demoscene (parties where people share prototypes and demonstrations that push the limits of computers as audiovisual tools); and a third in Southern Europe and the Iberian peninsula, organized around hackerspaces in squats and social centers and most closely linked with active extra-parliamentary left social movements. These RTCs all focus on providing infrastructure for activists: primarily email, web hosting, and chat servers. They are typically locally oriented and support activist individuals, groups, and networks in a particular city or country. He describes them as follows: "Radical technology collectives build political solidarity and nurture security behaviors within and between activist groups in addition to providing things like email and putting the right cryptographic algorithms in place."[54]

In addition, Maxigas notes that one of the most crucial functions of RTCs is maintenance and repair of movement infrastructures: "Even though the actual everyday practice of hacktivism is mostly about maintenance, groups that run infrastructure have received little to zero attention so far. This is especially ironic because even the emblematic movement of contemporary hacktivism (Anonymous) could not operate without relying on the services of radical server collectives. While it is the spectacular acts of disruption that go down in history, the daily labor of infrastructure maintenance makes history to a comparable degree. Therefore, it is necessary to rethink the history of technological resistance from a use-centric point of view in order to counterbalance innovation-centric narratives."[55] ICT maintenance and repair activities are rarely discussed in accounts of activist technology because of the mythology of innovation and the constant bias toward the "new," even though these activities are just as important to activist goals.[56]

Ultimately, A/I played a key role in bringing Italian activism online. The collective connected thousands of activists from a wide range of political backgrounds with their first email accounts, mailing lists, and websites. In the book, A/I members describe how they showed up everywhere: at protests, at fairs, at squats, and at meetings, convincing activists who at the time saw computers as something that were only used in the workplace or by the state that there was value in adopting these new tools for organizing and communications. A/I also maintained, supported, and repaired movement ICT infrastructure over more than a decade while serving as an important informal educational network for an entire generation of tech activists.[57]

RTCs have played similar key roles in nearly every region of the world over the past two decades. For example, in North America, RiseUp and May First/People Link perform the same kind of work as A/I for literally thousands of movement organizations.[58] It was from this social movement context that Open Whisper Systems, led by hackers from RiseUp, developed Signal secure messenger. This provides yet another example of technology design led by RTCs, deeply embedded in social movements, that then becomes industry standard: Open Whisper System's end-to-end encrypted messaging protocol was adopted in 2016 by WhatsApp, the largest messaging app in the world, with over 1.5 billion users.[59]

Design Scoping and Framing

One of the most powerful, and least discussed, ways that narratives structure design processes and outcomes is in the scoping stage: How do we frame the "problem?" Indeed, problem definition is a key component of all design processes. Herbert Simon, sociologist, economist, and author of *The Sciences of the Artificial*, argues that design always involves the recognition of assumptions and the redefinition of the design problem.[60] Yet much of the time, powerful institutions frame problems for designers to solve in ways that systematically invisibilize structural inequality, history, and community strategies of innovation, resilience, and organized resistance. In this section, I provide a critical reading of the design scoping process and highlight alternative approaches. I argue both for a shift from deficit- to asset-based approaches to design scoping, and for the formal inclusion of community members in design processes during the scoping or "defining our challenge" phase of a design cycle, not only during the "gathering ideas" or "testing our solutions" phases.

Scoping is ongoing, iterative, and fundamental to design work. It is sometimes thought of as a task to be completed before the "real work" of design can begin. However, design can also be seen as an iterative process of "changing the problem to find the solution." In his classic text on reflective practice, philosopher and urban planning professor Donald Schön notes that *problem framing* is one of the fundamental elements of design.[61] Schön grounds this idea in the work of philosopher, psychologist, and educational theorist John Dewey. In his theory of joint inquiry, Dewey argues that because knowledge is *particular* and *contingent,* rather than *universal* or *necessary,* "people jointly explore, discuss, and define a problem and jointly explore, develop, and evaluate possible solutions."[62] The way that a problem is conceived and framed has real implications for the range of possible solutions. Thus, for Dewey, determining the scope of a project is always a critical ethical decision.

In a related vein, designers and engineers Robert Hoffman, Axel Roesler, and Brian Moon note that many people believe that designers work systematically, in a top-down approach that starts with goals, requirements, and constraints, then moves toward solutions. However, they argue that research on how expert designers actually work shows that regular deviation from such a linear process is the norm.[63] Design proceeds through the alternating recognition and relaxation of assumptions, moving through iteration toward a "satisficing" solution: "The designer decides what constraints to relax in order to respond to the most important ones. The design concept that emerges from this process of sacrificing secondary properties is a satisficing design solution, not necessarily an optimal one, as is generally approached by engineering optimization. The satisficing solution is a necessity when trying to address a complex design problem with so many parameters that optimization approaches would not be feasible."[64] This view of a design process as ongoing problem iteration, concurrent with consideration of possible solutions, is shared by cognitive psychologists Linda Willis and Janet Kolodner, who refer to *design problem evolution* as the way that a designer "grapples with contradictions, ambiguities, and specification roadblocks and repeatedly reformulates the problem at hand."[65] Scoping is therefore an ongoing and key aspect of any design process. Unfortunately, under white supremacist capitalist heteropatriarchy and settler colonialism, scoping also often is used as an excuse to ignore, bracket, or sideline questions of structural, historical, institutional, and/or systemic inequality.

As noted HCI scholar Paul Dourish remarks, design often functions as an *antipolitics machine,* to use anthropologist James Ferguson's term for the depoliticizing effects of international development discourse:

> Development systematically forecloses an examination of the political contexts within which the development engagement takes place—the reasons for income disparity, the conditions of inward investment, the nature of democratic processes, the history of colonial relations, the effects of globalization, etc. Ferguson argues that the effectiveness of development projects are fundamentally constrained by the fact that the development discourse does not allow one to examine the conditions under which it arises. A similar argument could be made about design discourse, in which commitments to technological determinism and technosocial progress leave little room for the political and historical.[66]

Later in this chapter, we will explore several examples of how design discourse functions as an antipolitics machine in practice.

Ninety-Nine Problems, but We Frame Just One

In their 2016 book *Grassroots Innovation Movements*, STS scholars Adrian Smith, Mariano Fressoli, Dinesh Abrol, Elisa Arond, and Adrian Ely note that the concept of framing is used differently in social movement studies and in the sociology of technology.[67] For social movement scholars Robert Benford and David Snow, collective action frames are "sets of belief and meanings that inspire and legitimate the activities and campaigns of a social movement organization."[68] The creation and circulation of new frames is an important social movement activity, since frames enable and constrain action and shape the emergence of social movement identities. In science and technology studies, however, "technological frames consist of the shared problems, strategies, requirements, theories, knowledge, design criteria, exemplary artefacts, testing procedures and user practices that emerge through social interaction in groups. They help us to understand what social actors deem to be reasonable in choosing and developing a technology."[69] How, then, do problem frames depoliticize design processes?

For one, by decoupling "design problems" from structure, from history, and from consideration of systemic, persistent, ongoing forms of oppression under the matrix of domination. For example, the ways that algorithms are used by various actors and institutions to reproduce white supremacist capitalist heteropatriarchy and settler colonialism is reduced to a critique of "algorithmic bias." The ways that the media system symbolically annihilates most communities and forms of human experience while it produces neoliberal hetero- and cis-normative subjectivities; promotes anti-Blackness; and normalizes the logics of the prison industrial complex, settler colonialism, and empire are reduced to questions about representational equity according to a limited group of identity categories.

The organization of an entire arena of human endeavor—the design of new technologies—according to the logic of the matrix of domination, whereby designers, imagined users, values, affordances, ownership, governance, and other aspects of design are all set up to systematically reproduce white supremacist capitalist heteropatriarchy, both in process and outcome, is reduced to critiques of a few "quirky" examples of gender or racial bias: "Isn't it so messed up that digital camera algorithms thought Asian people had their eyes closed when the picture was taken? Whoever wrote that program is really racist!" Thus, the constant and pervasive reproduction of structural inequality across every field of life, including the design of digital tools, platforms, and systems, is reduced to individualized racist acts or, more typically, instances of unconscious bias.

This framing produces a particular set of conversations and actions. It allows technocentricity and solutionism to carry the day. If what we're up against is a particular software development team that made some bad decisions, then all we need to do is work to reengineer the software and smooth out biases that were accidentally coded in. Of course, code and UI bias reviews; algorithmic bias audits; antiracism workshops; gender parity targets for hiring, retention, and salaries; and increased awareness of microaggressions are all important and worthwhile pursuits. At the same time, if we never zoom out to the big picture, then we never take on the larger structures that constantly militate toward the reproduction of designed inequality. Instead, we are left to constantly put out tiny brush fires on our doorstep while the entire forest around us is consumed by a massive blaze. We remain forever stuck gathering donations of bottled drinking water for children in Flint, Michigan, without ever organizing to force the state to replace the contaminated water source and pipes throughout the entire system.[70] For these reasons, design justice as a framework impels us to reconsider the scoping process.

Problems with Problem Scoping: *The 18F Guide to Lean Product Design*

For example, consider *The 18F Guide to Lean Product Design* (18F is a federal office tasked with supporting other government agencies to build and improve tech products and services). In describing its design process, 18F notes: "The first stage of any project is to do research to discover problems that need solving. Your goal is three-fold: Identify and more deeply understand the challenge facing the organization and its stakeholders; Identify the people you believe could be most helped by your solution; and, Explore the problem, context, behaviors, and motivations of the people (your intended users)."[71] The guide then provides the following example:

> The challenge: the United States has high unemployment rate and the growth in jobs is for highly skilled workers. We need more citizens who can meet that demand, and we have evidence that college educated workers are more employed and more employable.
>
> The people: High school graduates and adults without a degree.
>
> The problem: Prospective college students lack information about the potential economic outcomes of a college degree, and also lack information that would lead them to be able to select which college is right for them.[72]

Design justice provides us with tools to critically analyze this problem statement. Recall that design justice centers analysis of how design affects the unequal distribution of benefits and burdens to groups of people at different locations within the matrix of domination.

The problem statement offered by 18F immediately skips over any discussion of structural inequality. For example: How is the unemployment rate distributed among different groups of people in the United States? How is college education distributed? What groups of people are getting access to those jobs that are growing, and what groups are being left out? In a design justice approach, the answers to these kinds of questions inform both the "people" and "problem" statements. Exploring these questions also modifies the assumptions undergirding the problem statement. The writers of this example universalize unemployment and access to college education across an unspecified "citizen," although both unemployment and college access in the United States are deeply structured by race, class, gender, disability, and immigration status: in other words, by location within the matrix of domination. The problem is framed as a *lack of information* about the utility of a college degree, rather than as any one of several alternative formulations that recognize intersectional structural inequality.

Recognition of racially disparate access to college would produce a different problem framing. For example, high schools that predominantly serve students of color often do not receive the resources they need to successfully prepare students for college; heavy policing inside schools and racially disparate application of disciplinary rules have led to a school-to-prison pipeline for low-income students of color; spiraling college costs have made higher education increasingly unattainable for poor and working-class students; and so on. The challenge, people, and problem, reframed through a design justice lens, shape a very different kind of design project—and a different allocation of resources, time, and energy.

Design Challenges: Full of Crap? Notes on the Gates Foundation's Reinvent the Toilet Challenge

Design challenges are a frequent, highly visible, and narrative-centric approach to design.[73] In 2011, the Gates Foundation launched a design challenge to develop a new kind of toilet. The rationale was as follows: "The Water, Sanitation & Hygiene program initiated the Reinvent the Toilet Challenge to bring sustainable sanitation solutions to the 2.5 billion people worldwide who don't have access to safe, affordable sanitation. Grants have been awarded to sixteen researchers around the

Figure 3.3
"Loughborough University has developed a user-friendly, fully operational household toilet system that transforms feces into biochar through the hydrothermal carbonization of fecal sludge." *Source:* Gates Foundation, n.d.

world who are using innovative approaches—based on fundamental engineering processes—for the safe and sustainable management of human waste."[74]

The goal of the challenge was to create a toilet that "removes germs from human waste and recovers valuable resources such as energy, clean water, and nutrients; Operates 'off the grid' without connections to water, sewer, or electrical lines; Costs less than US$.05 cents per user per day; Promotes sustainable and financially profitable sanitation services and businesses that operate in poor, urban settings; [and] Is a truly aspirational next-generation product that everyone will want to use—in developed as well as developing nations."[75] Between 2011 and 2018, the Gates Foundation invested more than $200 million USD in the Reinvent the Toilet Challenge and related toilet R&D.[76] Projects each received up to $100,000. Prototypes used high heat to convert feces into biochar, boiled black water to extract pure water, and added chemical agents to break down waste, among other technologies. In 2013, the foundation announced $5 million for Chinese researchers; in 2014, announced another $2 million for Indian researchers; and in 2018, held a Reinvented Toilet Expo in Beijing, where it announced a commitment of up to $200 million more, as well as $2.5 billion (billion, with a b, not a typo) in financing from the World Bank, Asian Development Bank, and African Development Bank. The challenge framing implied that the key problem is the failure of developing country cities to provide sewage infrastructure, combined with additional problems such as women's fear of using public toilet facilities in contexts in which they might be attacked by men.

However, sanitation experts working in developing countries argued that "the communities that desperately need sanitation will be unable to afford the advanced technology the initiative promotes."[77] For example, unrelated to (and ineligible for) the Gates Foundation's challenge, Toilets for People, a for-profit business focused on developing affordable toilets for developing countries, designed a low-tech waterless composting toilet with a cost below $200. A Swedish firm called Peepoople designed a biodegradable bag that kills pathogens.[78] Sasha Kramer, cofounder of an NGO that focuses on sanitation in Haiti, put it this way: "Building the toilet is the easy part. The most challenging step is making it work on the ground. The true challenge is not technology, it's really an issue of access, social mobilization, and ongoing maintenance of the toilet."[79]

Meanwhile, very low-cost alternatives to sewage infrastructure and large centralized treatment plants already exist and have been effectively integrated into daily life in a wide range of locations for decades (and in some cases millennia). A quick scan of DIY black- and greywater treatment technology on appropedia.org provides detailed descriptions of more than a dozen treatment approaches that can be built easily by individuals, families, or communities using locally available materials, at costs affordable to nearly all people on the planet.[80] For example, one of the most common and affordable solutions is a bucket toilet compost system. In essence, human waste is deposited in a bucket, then covered with a few scoops of ash, sawdust, or wood chips. When partly full, the bucket is then emptied into a compost bin. The compost bin must be regularly rotated. Within one year, the waste is converted to useful soil, which is safe to use for agricultural purposes. Larger-scale versions of this system are widely deployed. In China, a typical village refuse management system involves large cement tanks in which human waste is composted, later to be reused for agriculture.

My point here is not to argue that new, innovative toilet technology is not desirable, that it is not possible, or even that the Gates Foundation grants to the toilet innovation teams were a waste (pun intended!). Instead, I present the Reinvent the Toilet Challenge for analysis through a design justice lens: What story is told? How is the problem framed?

Who decides the scope? What values are built in to the designed objects and processes? Who benefits? Who loses?

To begin, the challenge ignored existing, low-cost, appropriate technology solutions to the "problem." Although tried and true, and arguably likely to be the most effective, these solutions were not mentioned in the challenge language, were not funded as recipients of challenge grants, and their uptake was not advanced at all by the billions of dollars linked to the Reinvent the Toilet Challenge. Jason Kass, in a well-argued *New York Times* op-ed, put it this way: "The trouble is that the Gates Foundation has treated the quest to find the proper solution as it would a cutting-edge project at Microsoft: lots of bells and whistles, sky-high budgets and engineers in elite institutions experimenting with the newest technologies, thousands of miles away from their clients."[81] This is not to say that existing systems are perfect, problem-free, or universally applicable. Sawdust requires cutting down trees, ash requires burning things and producing potentially harmful emissions, compost requires physical space for a composting unit, and so on. However, the complete lack of investment in thinking about how to improve already-existing solutions might reasonably be described as a grand failure. This failure is produced by a techno-solutionist orientation (new technology will save us!), an exclusionary and elitist understanding of what technology is and where it comes from (smart scientists in university and corporate laboratories), and a lack of interest in preexisting, community-based design practices.

Urvashi Prasad, writing about the Gates competition in 2012, notes that "we can't get distracted by the relative glamour of a technical design competition. Sadly, no perfect toilet for the poor will get us where we need to be. We also need an arsenal of non-technical strategies."[82] Prasad goes on to argue for optimization of solutions that already work, including community ownership of existing infrastructure and toilet blocks, enforceable contracts with private toilet operators, and flexible payment options for urban slum dwellers, such as monthly passes. They also note that in places where toilets are not in regular use, there is a wide range of reasons that people might buy in (or not) to their importance. For example, people may adopt toilet use as a sign of social status, for protection of female family members from having to engage in late night trips to communal toilets, and so on.

Prasad also highlights the importance of contextualized design: "Not even the best designed toilet technology will fit every situation. For instance, even in a well-established slum that has access to both water and sewage pipelines, individual homes may be extremely space constrained. Where exactly do you fit a new toilet in a 12 square metre—or 129 square foot—home that is sandwiched between three other equally compact homes?"[83] Ultimately, Prasad argues that uptake, use, and maintenance of existing solutions, as well as understanding real-world barriers and motivations, are the true keys to success. For example, they describe how the Slum Networking Project in Ahmedabad found that slum dwellers who decided that toilets were important to their community were willing to invest several times what the government contributed to maintain and upgrade sanitation infrastructure. Prasad urges that "those of us working to promote universal access to clean water and sanitation must keep our eyes not just on the competition and prizes, but on the less glamorous work of encouraging adoption, usage, and maintenance."[84] They also argue that existing public and private toilets make up about 50 percent of sanitation infrastructure in Indian urban slums, but many of them are poorly maintained. Figuring out how to promote ongoing maintenance, then, is the key challenge, rather than new toilet design. This is remarkably similar to Maxigas's argument for the important infrastructural and maintenance work of radical technology collectives, described earlier in this chapter.

Again, the point here is not that new technologies are useless, that design challenges are a waste of time, or that existing solutions are always sufficient. Instead, it is to recognize that wherever there are problems, those most affected have nearly always already developed solutions; that existing solutions that come from those most affected are likely to have the advantage of being based on local materials, skills, and infrastructure; that people who are from, and work directly with, the most affected communities should be included in and control design processes that are meant to benefit them; that sometimes (although not always) external resources can best be used to support, improve, scale, and/or reduce the costs of existing, locally created solutions; that barriers are often not about a particular tool or object, but are social, cultural, and economic in nature.

The Gates Challenge assumes the opposite of most of these points, at least in public discourse and in grantmaking: it ignores existing solutions; assumes that solutions will come from university labs far from the social realities of those without access to sanitation; makes no stipulation for, or even suggestion of, a codesign approach that would include local expertise and/or tacit and experiential knowledge; and makes no mention of adoption, usage, or maintenance.

The Loughborough University recipients of one of the Gates grants produced a very interesting article about their design process. According to them, design unfolded in phases, beginning with user research involving "multidisciplinary teams of experts" to gather requirements and focus groups with "primary users and secondary users," although they do not indicate whether the focus groups were actually conducted with Indian slum dwellers who lack access to toilets.[85] The team developed functional requirements, and the project hired an undergraduate student in the industrial design program to design a toilet seat. Next, the researchers and undergraduates took a field trip to India to meet with local toilet providers, aid agencies, and experts. Upon returning from the field trip, they created prototypes using blue foam and tested them with students and faculty from the design school. One of the key findings of the research phase was that "the product should be designed with local manufacture in mind, as this could be beneficial on a number of counts, from sustainability and cost through to ownership, maintenance, and repair."[86] I argue that this finding is generalizable to most design processes and should in fact be a starting point, rather than a conclusion.

The team also shares this fascinating finding: "Contrary to some popular beliefs in the West, a notable proportion of users (certainly in a domestic context) in Ahmedabad at least aspire to own a sit type toilet, despite the documented health benefits of squat toilets."[87] This presents a difficult moment for a design justice approach. If people everywhere aspire to own sit toilets, despite the health benefits of squat toilets, because their hegemonic presence in mass media and in the homes of local elites makes them a marker of economic success, what is the appropriate path? On the one hand, resources might be reallocated from product design to popular education and/or media campaigns about the benefits of squat toilets over sit toilets. On the other hand, design justice urges us to respect and support communities in making their own decisions. This is a perhaps unresolvable tension.

In any case, the Loughborough University team's integration of a biochar system seems to have happened entirely outside the user-facing design process: there was no prototyping process that involved real-world users (industrial design students created and tested foam prototypes after their short field trip), and the finished product was presented at the fair without ever having been tested in real-world conditions by real-world users. The paper ends with this statement: "Extensive user testing in the field will no doubt highlight issues that still need to be addressed, as well as possible flaws in the initial design."[88] No doubt.

In one of the most insightful articles about the challenge, Lloyd Alter writes that the winning projects are all expensive, complicated, and difficult to maintain.[89] They also require more household space than most intended users possess. Some are potentially deadly: several superheat feces, while others, like the California Institute of Technology (Caltech) winning entry, produce deadly chlorine gas. Alter also writes about the wasted water and energy involved in all flush toilet systems. Most interesting is his discussion of the history of humanure: "The fact is, you don't need high tech to deal with poop and pee, you need a social organization like they had in China and Japan before the development of artificial fertilizer. There was an entire economic infrastructure, like the boats and canals shown above in Shanghai, for picking the stuff up, processing and storing it to kill microorganisms, and using it as fertilizer. It was valuable stuff." He goes on to cite Kris De Decker's writing about the trade in human manure at the turn of the century, when the concession to manage collection, processing, and distribution from the city to the countryside was worth hundreds of thousands of dollars: "In 1908, a Chinese business man paid the city [the equivalent of $700,000] to obtain the right to remove 78,000 tonnes of humanure per year from a region of the city to sell it to the farmers in the countryside."[90] Humanure, historically, was a valuable commodity and an input to sustainable farming practices. Now replaced by imported fertilizers and phosphorous, it has been reframed as waste, and municipalities spend vast sums of money annually to literally throw away a potential source of income. Seen in the light of this history, the Gates Challenge might do better to invest in new businesses that purchase (or collect at no cost) humanure from slum dwellers for processing into fertilizer for sale to farmers, whether in urban farming or after transport to the countryside.

Finally, to highlight another one of the ways that the design narrative invisibilizes the matrix of domination, much writing in this field notes that people in India don't want to send their daughters to shared toilets because they fear they will be harassed.[91] If sexual harassment is one of the primary barriers to sanitation access, then it should follow that people interested in improving access to sanitation should invest in eradicating sexual harassment. However, this simple insight is entirely absent from the framing, scoping, prize eligibility, publicity, and the rest of the institutional narrative around the Reinvent the Toilet Challenge. It might make sense to perform a cost/benefit analysis of what it would take to eliminate (or drastically reduce) sexual harassment of women and girls during public toilet use versus what it would take to install functioning toilets in each household in India.

The boundaries of any given design narrative, such as in a design challenge, typically constrain the possibilities of addressing systemic issues, root causes, or approaches based on social organization. In the Reinvent the Toilet Challenge, low-cost existing technologies, business models that value humanure as a main input to fertilizer, and the need to eliminate or drastically reduce sexual violence against women and girls who attempt to use shared sanitation facilities are all off the table. This is true even though any one of these three approaches is potentially more likely than a new toilet design to achieve the stated goal of significantly reducing the proportion of the world's population that lacks access to clean water and sanitation.

Of course, new technologies are exciting and sometimes do bring significant improvements to quality of life and human capabilities, and any given design initiative hopes to maximize impact by tightly focusing on a particular aspect of a broader puzzle. That said, design challenges constructed with little to no input from the most affected people, that assume that solutions will come from university experts thousands of miles away, that ignore existing solutions, and that systematically avoid the root causes of identified problems are not grounded in design justice, and ultimately they are likely to fail in both practical and ethical terms.

Happily, there are a growing number of people, organizations, and networks that recognize these points and are working directly with communities with lived experience of design challenge areas to frame, scope, prototype, and do design work together. This is taking place in design narrative workshops in the Design Justice Network, in spaces like the Make the Breast Pump Not Suck Hackathon and Policy Summit,[92] MigraHack,[93] and Trans*H4CK,[94] and elsewhere throughout multiple design fields. In chapter 4, we will dive more deeply into the question of how to organize hackathons, DiscoTechs, and other spaces for technology design in ways that challenge, rather than reproduce, the matrix of domination.

Design Narratives: Conclusions

Stories have power. The "official" Twitter origin story holds that one of the founders had a brilliant blue-sky flash of genius. Developers who were part of the process have a counternarrative: anarchist hacker-activists created TXTmob as a tool to help affinity groups stay one move ahead of the cops in the NYC Republican National Convention protests of 2004, and TXTmob served as demo design for the Odeo hackday that led to Twitter. The key point is that the stories we tell about the design of new technologies both reflect our broader understanding of the world and shape the horizon of the possible.

Design generates attention, and attention is an increasingly scarce resource that is not equitably allocated. The amount of attention we can command is shaped, in part, by our location within the matrix of domination (white supremacy, heteropatriarchy, capitalism, and settler colonialism). A design justice approach requires proper attribution for too-often-erased participants in design processes.

Innovation in media technology, like all technological innovation, is an interplay among complex sets of actors including users, developers, firms, universities, the state, and others, not a top-down process led by solitary programmer "rock stars." Lead users develop many, if not most, innovations in any given field, through DIY and informal processes outside of "official" research, design, and development channels. This has implications for the way we think and talk about design, as well as policy implications. Social movements in particular have always been a hotbed of innovation in media tools and practices, in part because of the relationship between the media industries and social movement (mis)representation. Social movements, especially when led by marginalized communities, are systematically ignored and misrepresented in the mass media, so they often form strong community media practices, create active counterpublics, and develop media innovations out of necessity.[95] Social movements thus can be important sites of technology design, diffusion, adoption, and support. Many social movement media innovations are later adopted by the journalism profession and by the broader cultural industries, although stripped of their original counterhegemonic intent. Examples include TXTMob and Twitter, Signal and WhatsApp, and many more. We have to tell these stories so that social movement contributions to the history of design are not erased.

Finally, one of the most important ways that narratives structure design is in the scoping and framing of design problems. Design scoping processes that exclude structural problems, large institutional actors, or the state from the field of analysis convert design into an antipolitics machine. Design narratives too frequently invisibilize the matrix of domination and set the boundaries of the imagination to exclude already existing, community-led solutions, as in the Gates Foundation's Reinvent the Toilet Challenge.

Design justice provides a lens that we can use to analyze design narratives. In other words, what stories are told about design problems, solutions, contexts, and outcomes? Who gets to tell these stories? Who participates, who benefits, and who is harmed?

Design justice considers a dual pragmatic/utopic approach that simultaneously offers concrete suggestions for immediate implementation to improve people's quality of life while also calling out power inequalities and larger structural forces that impact people's life chances in the long run. Design justice also approaches scoping and framing through a community asset lens, and recognizes that communities that are marginalized under the matrix of domination nearly always have already developed strategies and tools to navigate their problems, as well as rich repertoires of sociotechnical practices to support cultural, political, and economic life. Design justice is interested in telling stories that amplify, lift up, and make visible existing community-based design solutions, practices, and practitioners.

4 Design Sites: Hackerspaces, Fablabs, Hackathons, and DiscoTechs

Figure 4.1
Cover of *DiscoTech* zine, by the Detroit Digital Justice Coalition.

Be excellent to each other, dudes.

—Noisebridge's One Rule[1]

In many ways 'hackerspace' is an elitist name for middle-class white guys screwing around with computers and making a big deal out of it. Come on. Every other block in this town has an auto body shop where more hacking takes place than y'all can imagine, and people have their own networks of friends and family and colleagues who learn stuff and create things. Nobody's writing about that in Wired. That has to set off your bullshit detector a little.

—Liz Henry, "The Rise of Feminist Hackerspaces and How to Make Your Own"

The acrid smell of hot solder emanates from a table in the corner, where an intergenerational group of people is learning how to build a pirate FM radio station. Across the room, at the beat-making collaboration station, three teenagers with headphones on nod in time to the *boom boom bap* of hip hop beats they are creating. In another corner, several children goof around in front of a giant green screen, where they shoot still images that they will later stitch together into an animation. These activities, and many more, are part of a DiscoTech, or Discovering Technology community fair, within the Media a Go Go Lab at the 2012 Allied Media Conference (AMC).[2] The description of that space in the AMC program reads: "Lab participants will learn DIY media making skills, collaborative design, innovative communications tactics, and build technology (transmitters, device controllers, etc.) throughout the weekend. We will create opportunities to analyze, remix, and transform our current and future media and technologies! This dynamic space is where we put the Walk to the Talk at the AMC."[3] Over the course of three days, hundreds of people will participate. Some come for focused workshops led by talented facilitators, others to "hang out, mess around, and geek out."[4]

DiscoTechs were first created a decade ago by community technologist Diana Nucera and the Detroit Digital Justice Coalition (DDJC). According to the DDJC, a *DiscoTech* is "a replicable model for a multimedia, mobile neighborhood workshop fair."[5] The first DiscoTech took place on Saturday, December 12, 2009, at the 5E Gallery in Detroit; it featured hands-on workshops about the Internet, electronics, public policy, and the growing community of organizations that had then recently linked to form the DDJC. There were consultation stations, electronics workshops, and film screenings. Attendees included seniors, youth, environmental justice activists, hip hop artists and producers, people on welfare, community organizers, artists, technologists, and others. Hands-on activities ranged from how to set up an internet account to how to build a computer using recycled parts. DiscoTech attendees also participated in interviews and surveys that were used to develop ideas for DDJC programs, many of which were later implemented using the federal Broadband Technology Opportunity Program (BTOP) funds that the DDJC successfully bid for in partnership with the University of Michigan.

The DiscoTech model spread widely after it was shared at the DiscoTechs Unite session at the 2012 AMC. There, the DDJC joined Broadband Bridge (a Washington, DC-based initiative that had also begun organizing DiscoTechs during the previous two years) and members of the AMP network to create the first AMC DiscoTech.[6] On Sunday, July 1, they invited conferencegoers to "come and jump around from station to station, discovering technology with your peers. It will be fun!"[7] This DiscoTech featured dedicated *collaboration stations* where participants could learn and practice design and technology skills focused on electricity, audio recording and beat making, soldering, mesh networking, and cryptography. The *DiscoTech* zine (figure 4.1) was also published in 2012 and distributed widely at the AMC and beyond.[8]

By 2013, Diana Nucera and Janel Yamashiro from AMP, Nina Bianchi from the Work Department (theworkdept.com), Andy Gunn from the Open Technology Institute, and I worked to expand the AMC DiscoTech into an ongoing three-day-long dedicated space. That year, the Discovering Technology Lab focused on DIY and do-it-together (DIT) technologies, collaborative design, making, hacking, hardware, software, and sustainable technologies. All attendees received a *DiscoTech* zine and were invited to "get down with Phunky Phone Phreaks, Fierce Fashionistas, Data Viz Wiz Kids, Documentation Doctors, Tech Help Desk Divas and our Wonderful Webmaking Friends!"[9]

Inspired by the AMC DiscoTechs, I connected with others to help organize similar events in the Boston area and elsewhere. In 2014, students and staff at the MIT Codesign Studio partnered with local organizations around the world to coordinate a series of "countersurveillance DiscoTechs" in Cambridge (Massachusetts), San Francisco, Ramallah,

Mexico City, Bangalore, and New York City.[10] I wasn't the only one excited by the model: also in 2014, the Bento Miso Collaborative Workshop hosted a design-focused DiscoTech in Toronto, featuring poster design, screen printing, bookbinding, comics, and stop-motion animation;[11] this event was a fundraiser for the Future Design Lab at the 2014 AMC. The 2014 Internet Governance Forum in Istanbul featured a DiscoTech organized by the Association for Progressive Communications, Tactical Tech, and the World Wide Web Foundation.[12] In 2016, the MIT Codesign Studio team, Research Action Design, Intelligent Mischief, and the DCTP supported local organizers in multiple cities to run Cooperative Economy Discovering Technology fairs (co-op DiscoTechs). These focused on the use of technology to strengthen worker-owned cooperatives, consumer cooperatives, housing cooperatives, and other aspects of the cooperative economy. The response was tremendous, with events in Boston, Salem, New York City, Boulder, Philadelphia, Oakland, and London.[13]

DiscoTechs continue to spread. They provide one excellent model for how to organize inclusive, community-centric events focused on participatory design, digital media, and technology. There are many other kinds of design events, gatherings, and spaces, such as hacklabs, makerspaces, fablabs, and hackathons. This chapter explores the question, "How do we apply design justice principles to create inclusive design sites?"

Design takes place everywhere, but particular sites are valorized as ideal-type locations for design practices. There is a growing literature about hacklabs, hackerspaces, makerspaces, and fablabs (various types of spaces where people gather to learn how to hack, make, and build), as well as about temporary design- and technology-focused events, such as hackathons. Unfortunately for design justice practitioners, this literature reveals a long-term shift away from hacklabs and hackerspaces as explicitly politicized spaces at the intersection of social movement networks and geek communities.[14] Instead, start-up culture and a neoliberal discourse of individual technical mastery and entrepreneurial citizenship have largely come to dominate hackerspaces,[15] even as city administrators have leveraged the popularity of technological solutionism to create municipal "innovation labs."

At the same time, we should not allow neoliberal discourse about these sites to erase their past, present, and future radical possibilities. There is a deep history, or alternative genealogy, of hacklabs and media/tech convergence centers as spaces tied to social movements. There has also been a recent move toward the intentional diversification of hacker and makerspaces, specifically along lines of gender and, to a lesser degree, race. Examples of this trend include Liberating Ourselves Locally, Double Union, and a wave of new, explicitly intersectional feminist spaces dedicated to hacking, making, crafting, and design. In this chapter, I argue that in addition to the diversification of hacklab participants, design justice requires a broader cultural shift in how such sites are organized. In particular, design justice implies an intentional relinkage of design sites to social movement networks.

Just as dedicated design sites need to be transformed, we must also interrogate the ideals, discourse, and practices of design events like hackathons. There is growing interest in reimagining design events to be more intentionally liberatory and inclusive, as in DiscoTechs, Occupy Data hackathons, MigraHack, Trans*H4CK, and the Make the Breast Pump Not Suck Hackathon and Policy Summit. This chapter critically engages the literature about, and real-world practices within, hackerspaces; traces the cooptation of hacker culture by neoliberalism; attempts to imagine more intentionally liberatory and inclusive sites where design justice principles and practices can be implemented; and describes the ongoing spread of intersectional feminist design sites. It concludes with specific recommendations for how to develop design sites that are informed by design justice principles.

Hack, Make, Hustle: Subaltern Design Sites, Marginalized Design Practices

Privileged design sites like hackerspaces, makerspaces, and hackathons are not the only game in town; indeed, most design takes place elsewhere. Oppressed and marginalized peoples already have their own design sites, practices, and communities, although these are often ignored, pushed to the side, made invisible, or made to seem "less important." What I call *subaltern design sites* have always existed.

As the authors assembled by science and technology scholars Alondra Nelson, Thuy Linh Nguyen Tu, and Alicia Headlam Hines, in their edited volume *Technicolor: Race, Technology, and Everyday Life* (2001) reminds us, subaltern design sites may be focused on normatively "high-tech" tools and practices such as computers and software development, but also may focus on "everday" technologies—for example, in auto workshops, cell phone repair shops, or in audio stores and sound system culture.[16] Car culture requires highly technical skills and design capacities; think about the work that goes into designing and maintaining lowriders.[17] Or consider the extensive scholarship on the history of Jamaican sound systems, the influence of the Caribbean diaspora on sociotechnical knowledge and practices that gave birth to hip hop in New York City (for example, the influence of Jamaican-born DJ and audio innovator Kool Herc),[18] and the appropriation of vinyl records and turntable technology to create a new, world-changing musical genre and cultural movement.[19] King Tubby, Lee "Scratch" Perry, and other Jamaican studio innovators created music recording techniques that now permeate all of global popular music, such as the *drop* (an approach to musical composition and the creation and resolution of rhythmic tension through the subtraction and addition of prerecorded tracks). They were also brilliant hardware hackers; for example, they created tape-delay effects by physically stretching a loop of magnetic tape around the studio, cut to the length that would produce triplets timed against the track's main rhythm, among many other techniques.[20]

At the same time, design justice also recognizes the importance of sites where people focus on design practices that have been raced, gendered feminine, and/or otherwise coded as less valuable or not recognizable as "technology." In some cases, women, femmes, and other oppressed people's design practices operate within microsites such as the home. Under conditions where only certain kinds of technologies, sociotechnical knowledge, design practices, and skills are recognized and promoted by larger institutional, cultural, political, and economic regimes, many design practices never receive resources or recognition. Another way to look at it is that design justice recognizes that many important sociotechnical practices are designed, developed, and shared through constant, small-scale interactions within the space of the home, the family, kinship networks, and within communities.

For example, consider the development and exchange of agricultural knowledge and technologies that takes place in sites such as village farms or community gardens.[21] Or to take another example, communication scholar Aisha Durham created the term *hip hop feminism* to describe both the social history of and the specific forms of sociotechnical innovation by Black women that produced a sea change in feminist organizing practices, such as the outreach and media strategies they used to organize the successful Million Woman March in Philadelphia on October 25, 1997.[22] There is a large and rapidly growing body of scholarly work that centers and recovers stories of sociotechnical innovation from the margins. Still, design practices, spaces, networks, and histories that are about women and femmes, QTPOC, and/or Disabled people remain marginalized, invisibilized, and under-resourced.

Invisibilized design practices also take place on the margins within larger institutions. Marginalized people working within institutions that they do not control often create in-group support networks that include sharing a wide range of knowledge and practices, including design skills. For example, Jose Gomez-Marquez, the codirector of MIT's Little Devices Lab, found that between 1900 and 1947, nurses (who were mostly women) not only constantly designed and modified medical technologies to improve patient care, but also shared and published their medical device innovations in a magazine called *American Journal of Nursing* (*AJN*).[23] Gomez-Marquez, inspired by this history of subaltern design practice, has been working with nurses and hospitals to open *MakerNurse* sites meant to support, facilitate, and valorize present day nurses' medical technology design knowledge, practices, and objects.[24]

Design, maker, and hacker cultures that originate in working-class communities, center women and femmes, and/or are based in communities of color don't receive the resources, visibility, validation, and respect that those centered on white, cisgender, heterosexual men do. These communities have deep but less-recognized histories of hacking, making, design, and innovation. This includes what mainstream economists refer to as *business process innovation*, as well as what the design community refers to as both *product* and *service design*. Service design innovations in working-class communities aren't necessarily referred to as *service design innovations*. Instead, people might use their own terms—for example, a side gig or a hustle, as described by media scholar

S. Craig Watkins in his new book about how these types of activities form an increasingly important part of the innovation economy.[25]

At the same time, invisibility may be strategic: subaltern communities sometimes shield their practices and innovations from mainstream visibility to avoid incorporation and appropriation. In addition, innovations in many fields often operate in legal grey zones, and systematically unequal policing may expose subaltern innovators to harm from the various arms of the prison industrial complex. This is most starkly visible in the United States today in the legalization of marijuana. After decades of a so-called drug war that saw hundreds of thousands of people, disproportionately Black and Latinx, incarcerated for marijuana use, possession, and sales, suddenly (primarily) white-owned companies are swooping in to capture the lion's share of the newly legalized marijuana market. Most of these companies participate in the discourse of technological innovation as they jostle to offer the "market-leading" app for on-demand marijuana delivery and to secure millions of dollars in venture capital funding.[26]

In addition, design justice practitioners recognize that neither subaltern design sites nor privileged design sites are utopias. Many, or most, of the power dynamics that we would like to critique and transform in the latter also often operate within the former. For example, an auto workshop may be a site for the development, expression, and sharing of sociotechnological knowledge and skills between working-class men while simultaneously reproducing heteropatriarchal norms of gendered technical knowledge and skills that exclude women and femmes. Or it may be a site where those norms are challenged or transformed. Similarly, a fashion design studio may be a site where highly technical knowledge about apparel design and production is developed and shared and may (or may not) be an inclusive space along lines of race, gender identity, and/or sexual orientation. The same site may also be a key node in capitalist relations of production and consumption, as clothing designers labor to create innovative patterns that are then produced in sweatshops by migrant workers who typically face long hours, low wages, abusive bosses, health hazards, and humiliating work conditions.[27] Nor does design justice ignore the ways that community, local, diasporic, and/or Indigenous design sites may sometimes be locations for sustained resistance to cultural erasure through the ongoing production of sociotechnical knowledge and designed objects, even as they may simultaneously reproduce heteropatriarchal values and norms that were often imposed through settler colonialism.

Design justice emphasizes the value of local, community, diasporic, and Indigenous knowledge, practices, design processes, and technologies. These have often been appropriated, undermined, attacked, and marginalized for centuries under colonialism and capitalism, but they have not been erased. Indeed, the history of capitalism is in large part a history of the extraction of design practices that once took place in family and community microsites and their subsequent systematization, rationalization, and modification to fit the requirements of mass production. For example, consider the transformation of agriculture from Indigenous knowledge of small-scale planting, harvesting, and land-management techniques to modern agribusiness with monocultures, pesticides, fertilizers, and roboticized megafarms;[28] the transformation of healing from women's work to modern medical science (accomplished only with great violence to women healers, as autonomist Marxist feminist scholar Silvia Federici documents in her brilliant and disturbing text *Caliban and the Witch*);[29] or the archetypal birth of the capitalist mode of production in the transformation of clothing from a home-based practice of design, production, and constant repair to a globalized megaindustry of sweatshop labor, fast fashion, and disposability.[30]

What, then, does a design justice approach have to tell us about privileged design sites?

Design Spaces: Hacklabs, Makerspaces, and Fablabs

Design justice is a community of practice that locates itself within longer social movement histories. For example, the book *Grassroots Innovation Movements*[31] (mentioned in chapter 3) describes six case studies: the UK movement for socially useful production, the South American appropriate technology movement, the Indian People's Science Movement (PSM), hackerspaces, fablabs, and makerspaces around the world, the Brazilian Social Technology Network, and the Indian Honey Bee Network. The authors contextualize hackerspaces, fablabs, and makerspaces within "a tradition of thought in modern environmentalism and development concerning accessible tools for local, sustainable developments ... that includes the social ecology of Murray Bookchin, Stewart Brand and the Whole Earth project, E. F. Schumacher's *appropriate technology*, Ivan Illich's *convivial tools*, alternative technologists such as Peter Harper and Godfrey Boyle, and ideas by Mike Cooley and others concerning socially useful production."[32] However, much of this history is erased by popular narratives of design and sociotechnical innovation.

Many are working to challenge that erasure. Media scholar Maxigas has carefully traced the evolution of hacklabs from key nodes in a global autonomist network in the 1990s to their more common present-day configuration as hacker playpens integrated into the neoliberal city.[33] This transformation was also described by designers Johannes Grenzfurthner and Frank Apunkt Schneider in their *Hacking the Spaces* zine for the Critical Making publication series.[34] They describe the origin of hacklabs as spaces for the micropolitical practice of alternate life pathways that emerged in parallel to the ascendant regime of capitalist globalization in the wake of the collapse of the vague utopics of late 1960s counterculture. In place of drugged-out, sloganized imaginaries of global revolution, they argue, some participants in the counterculture shifted their energies to the creation of concrete alternative communities. This involved various projects focused on building the new world in the shell of the old, rather than attempting to institute systems transformation from above by seizing state power. Grenzfurthner and Schneider's reading of this history leaves race and gender unmarked, despite the fact that many autonomist practices were directly inspired by ongoing dialogue with Black Marxists like C. L. R. James, by the Black Panther Party's free breakfast and education programs, and by Black women in the wages for housework movement.[35] As they put it: "The autonomia movement of the late 1970s that came to life in Italy and later influenced people in German-speaking countries and the Netherlands was about appropriation of spaces, be it for autonomous youth centres or appropriation of the airwaves for pirate radio. Thus, the first hackerspaces fit best into a countercultural topography consisting of squat houses, alternative cafes, farming cooperatives, collectively run businesses, communes, non-authoritarian childcare centres, and so on."[36] The authors argue that participants in this first wave of hackerspaces were explicitly antiauthoritarian and opposed both capitalism and authoritarian communism. They also rejected bourgeois norms, culture, values, and lifestyles. Often physically located within squats, these hackerspaces served as models for an alternative spatial organization of life because they were mixed environments for work, play, and sleep. However, as they note, "alternative spaces and forms of living provided interesting ideas that could be milked and marketed. So certain structural features of these 'indie' movement outputs were suddenly highly acclaimed, applied and copy-pasted into capitalist developing laboratories."[37]

Communication scholar Fred Turner describes a closely related dynamic in the United States, where he traces the cultural origins of Silicon Valley–style libertarian techno-utopianism to failed California communes. Turner also discusses the influence of the Burning Man festival on the rise of Silicon Valley and the information economy. For Turner, capitalism is endlessly adaptable and uses the energy and fresh ideas of the counterculture to revitalize itself.[38] Grenzfurthner and Schneider argue that something similar took place with hackerspaces in the European context, which they describe as originally being "third spaces" outside of the logic of both the communist state and the capitalist market. Initially, people were drawn to these spaces as highly politicized countercultural communities where life, work, and play could be seamlessly blended. However, the authors argue that, ultimately, many ecological countercultural ideas and projects turned into trendy "green" or "sustainable" businesses, which provide a reservoir of positive affect for continued participation in the capitalist system.

A less totalizing narrative of this process might be that radical ideas and practices that were pioneered by people working within antiauthoritarian social movement networks were, in some cases, adopted by corporate actors and thereby scaled up and normalized. In other words, another reading is that anarchist ideas and, in some cases, individuals were able to infiltrate capitalist institutions, and through technical systems design, they spread certain kinds of decentralized power throughout society (e.g., in internet architecture).

In any case, the transformation of hackerspaces from radical nodes in autonomist movement networks to geeky havens geared toward sprouting new start-ups took place in the long context of the end of the Cold War, the collapse of Communist states, and the heady, mythological moment of the global triumph of liberal democracy and neoliberal markets.[39] As city governments reconfigured themselves for the age of free market triumphalism, and as they "sanitized" urban cores to attract high-skill information industry jobs, reverse the process of white flight, build tourist economies, and gentrify, they also cracked down on and closed most of the squats. A few were converted into loft spaces for (mostly) white urban bohemians, "creative workers," and hipsters.

As recently narrated in compelling detail in the book *Kaos: Ten Years of Hacking and Media Activism* (2017), a page-turning, collectively authored history of the radical Italian tech collectives Autistici/Inventati, squats and social centers linked to the antiauthoritarian left were key sites for the European hacker activist scene in the 1990s and into the 2000s. Many, but not all, of these spaces were later evicted by police. They were dismantled or pushed out in the drive toward redevelopment of the city centers for tourism, revitalization, and the creation of sanitized innovation hubs or entrepreneurial zones.[40] Hacklabs were thus transformed from semi-permanent social anarchist enclaves into sites for the production of neoliberal entrepreneurial subjectivity.

Hacklabs in the Global South

Even as Maxigas, Grenzfurthner and Schneider, Toupin, and others provide thoughtful and critical histories of hackerspaces and hacklabs, and critique such spaces for their recent depoliticization and the ways they often unwittingly reproduce patriarchy and racism, they also largely ignore the rich history of hackerspaces outside of the European and US contexts. The Latin American hacker scene, for example, is largely invisible in their accounts. Digital media scholar Andres Lombana Bermúdez has written in depth about Latin American hacker- and makerspaces.[41] For example, there are Territorial Innovation Centers (*Laboratorios de Innovación Territorial*) in Colombia, as well as "make, tinker, and learn" creative camps in several locations in Central America. TecnoX is a growing network of open hardware hackers from across Latin America who are increasingly visible and engaged in conversations about how to connect open hardware hacking to social movements.[42] Brazil-based transfeminist hacker organization Coding Rights, led by lawyer and technologist Joana Varon, uses research, prototyping, design, and meme culture to challenge data colonialism, gender-based violence, and structural information inequality across the Americas (codingrights.org).

In Cuba, media and culture scholar Paloma Duong describes DIY neighborhood networks organized by gamers, as well as the *paquetes*, or sneakernet content-delivery networks organized by entrepreneurs who physically distribute copies of films, music, and games via USB drives.[43] Anthropologist Sujatha Fernandes explores the ways that urban social movements in Venezuela pushed the state to support community control of ICTs through a change in telecommunications law to allow community radio and TV stations, as well as through the establishment of ministries of popular telecommunications and popular information with multimillion dollar budgets. She also notes the tensions between the state apparatus and the movement organizations along the way.[44]

The Instituto de Midia Etnica in Salvador, Brazil, founded by Paulo Rogerio, has an Afrocentric media, tech, and design center called *Ujamaa*. The site boasts a hacklab, a guest room for visitors, a kitchen, and a space for talks and events. Ujamaa regularly hosts design workshops, such as the *Ocupação Afro Futurista* ("Afrofuturist Occupation"), and works to raise awareness of the long history of Afro-Brazilian sociotechnical innovation.[45] Also in Brazil, under the Workers' Party government of Lula Inacio da Silva, Minister of Culture Gilberto Gil promoted and supported a network of *Puntos de Cultura*, or cultural hotspots. These community media centers, powered by free software, provided infrastructure for cultural production and circulation in low-income neighborhoods throughout the country. This experience was also replicated in Argentina.[46] Puntos de Cultura became sites where neighborhood youth developed digital skills such as music recording and editing, video production, graphic design, and web development.

Communication professor Anita Say Chan, in her book *Networking Peripheries*, provides a powerful overview of the ways that technological innovation often happens on the margins of society, far from the innovation hubs imagined and created by city planners, state officials, and private sector investors. She describes how digital cultures that emerge organically from the peripheries, including in the Global South, are different from those produced via the universalizing imaginary of technosolutionists who operate from positions of great power.[47]

It's beyond the scope of this chapter to trace the scale of too-often invisibilized regional and local design sites across the entire Global South. Other scholars and practitioners, such as those cited here (and many more, such as those in the Decolonising Design group at decolonisingdesign.com), are already doing this work. I hope that over time more inclusive histories of design sites will emerge. This will help us envision the possible liberatory futures of design justice sites in ways that are global in vision and aspiration, while also deeply rooted in local and regional specificities.

Hacklabs in the Neoliberal City and the Rise of Innovation Hubs

Over the last three decades, the private sector, the academy, and the state (in that order) all recognized the power of hacklabs and moved to incorporate them into their respective innovation strategies. Innovation labs are increasingly popular at private universities; for example, the Annenberg Innovation Lab at USC, the Harvard Innovation Lab, and so on. Cities everywhere, like universities, are also setting up innovation labs. Boston is home to the Mayor's Office of New Urban Mechanics; in Los Angeles, there is the Civic Innovation Lab (CIL). The CIL frames itself as follows: "Part design lab, part community caucus, part accelerator of urban solutions, Civic Innovation Lab at Hub LA is dedicated to the development of real solutions designed with and for communities throughout Los Angeles."[48] The CIL launched with a design event in September of 2014, for which people were encouraged to come up with possible design challenges. The next move was a call for solutions, followed by a selection of projects to be "incubated" on the path to becoming start-ups. We can read this is a move in the right direction: city government is becoming more transparent, more administrative data is being made available to the public, and administrators are actively seeking ideas from engaged publics about how to improve government services and city residents' quality of life. Through such initiatives, city administrators signal that the public is invited to participate in decisionmaking on an ongoing basis, not only at the ballot box every few years. Innovation labs, through this lens, move us toward the everyday practices of participatory democracy and have the potential to include many more people in the design of city systems. It is also encouraging that residents were included in the call to define the design challenges because, as noted in chapter 3, power over framing and scoping design challenges is so important.

More cautiously, we might say that these developments are positive, but imperfect. First, the "stakeholders" involved in these processes are typically not representative of city residents. In most public participation design processes, unfortunately, research shows that elite participation (by class, race, gender, education, language, and so on) is the norm.[49] Accordingly, design challenges and solutions are typically limited in scope to elite concerns. In addition, implementation is key: city innovation labs may develop excellent ideas and prototypes, but without top-down buy-in from city officials, department heads, technocrats, and administrators, as well as bottom-up buy-in from community-based organizations, adoption often fails.

It's also possible to read municipal innovation labs more critically within the larger context of city officials, planners, and the real estate industry collaborating to rebrand cities and attract tech companies by establishing innovation centers, hubs, and/or zones. For example, Boston created an innovation zone;[50] Cambridge, Massachusetts, and MIT are developing a software, aerospace, and biotech development zone around Kendall Square;[51] New York City partnered with Cornell University to transform Roosevelt Island into an innovation zone;[52] Los Angeles provided incentives for tech firms to establish offices along the Los Angeles River in Downtown Los Angeles; and so on.

In a 2015 article, civic technology professor Eric Gordon and graduate student Stephen Walter trace a brief history of the rise of innovation offices in American cities.[53] They argue that the growth of urban data systems begins with New York City's CompStat crime database, designed for internal use by the NYPD, then spreads to other city offices (such as ParkStat and HealthStat) and is later adopted by other cities: Baltimore, Boston, Los Angeles, and more. They also discuss the creation and spread of the 311 system, designed to capture citizen feedback via voice calls. As they describe it, the Obama administration's open data directive pushed federal agencies to make data available, and also served as a catalyst for many cities to open data sets, as well as to create and maintain application programming interfaces (APIs) to allow both individuals and private companies to build new services on top of public data. Along the way, the authors note the steady growth of high-level city administrative positions such as chief digital officer and chief innovation officer. These positions are often filled by individuals who come from the private sector. People with backgrounds in technology start-ups and internet companies bring the language, design approaches, and values of the for-profit sector into city government and promise to use their experience to make city government more user-friendly and efficient.

On the one hand, this process does produce an improvement in the usability of many city service interfaces. At the same time, citizens become conflated with users—and while users simply take actions within a framework, citizens participate in constructing that framework.[54] Neoliberal, technocentric ideas about the city as a machine or as a software system waiting to be optimized have become increasingly prominent. I agree with these authors, and with digital media theorist Wendy Chun, that citizens should not be reduced to users through the lens of neoliberal governmentality.[55] At the same time, I believe "users" can also be reconceived as active participants in the design and (re)production of technologies (as discussed in chapter 3).

In addition, when design sites emerge organically, they are not islands: they are hubs within thick networks of practitioners or gathering places for vibrant cultural scenes. This is why top-down innovation spaces that are planned and built by powerful institutional actors like city governments or private companies often feel forced. They don't emerge from an existing, dedicated community of practitioners; they usually don't reflect local specificity, culture, and assets. Rather, they draw on globalized, universalized, abstracted ideals about what constitutes innovation. Typically, by default they encode and reinscribe raced, classed, gendered, ableist assumptions about innovation, design, and creative industries. They reproduce what Arturo Escobar describes as the *one world ontology* instead of serving as sites for the production of pluriversal possibilities.[56]

This happens at multiple levels: spatial, aesthetic, discursive, and linguistic, as well as in terms of membership, staffing, governance, resource allocation, and so on. For example, consider the location of an innovation space in a particular kind of neighborhood in the city; the type of building it's located in; the aesthetics of the space itself. Consider the ways that the space is talked about, who it is meant to serve, and the visual and written style of the propaganda about the space. Think about the way that such spaces are typically monolingual in the dominant language of the nation state, culturally geared toward the dominant racial/ethnic group, and discursively and pragmatically gendered. For example, how many innovation spaces include translation services? How many include childcare? Rooms for pumping breast milk?

Given both broader social structural inequality and the reproduction of those structures in the ways just described here (spatial, aesthetic, discursive, cultural, political economic, etc.), membership in these spaces also tends to skew heavily white, middle-class, cisgender, and male. Governance (decision-making about how the space operates, how to allocate the resources that it attracts, what the priorities are, etc.) is typically dominated by the same group of people.

Innovation offices also tend to reproduce neoliberal values of efficiency, predictability, and individualism. The individual user replaces both the citizen and the community, not to mention the community-based organization or the urban social movement. The limits of citizen action, as imagined by tech-sector transplants to city government, are constrained by "good" citizen behavior and largely center on the "happy customer" who has a pleasant interaction with new, streamlined city services (making appointments at the DMV, paying parking tickets, etc.). The good citizen in the neoliberal city is also imagined as a contributor to public reporting systems set up and maintained by city administrators, most likely via a contract with a private-sector, for-profit firm (such as Textizen). In this way, "city government is masking its authority under this promise of collaboration as it redoubles its hold on power by dispersing it to the governed."[57] Innovation offices create, maintain, and promote platforms that facilitate the offloading of tasks traditionally performed by government onto city residents.

Just as users provide free labor for the dominant platforms in the cultural economy, neoliberal citizens provide free labor for city managers on the dominant urban incident reporting platforms. Citizens are encouraged to report potholes, petty crime, and graffiti and are rewarded with promises of more rapid service delivery. Journalism professor Michael Schudson describes these practices as *monitorial citizenship*.[58] Monitorial citizenship can be read as a key part of the privatization of public services, the conflation of the citizen with the user, and the spread of neoliberal governmentality into administrative discourse. This process also produces neoliberal subjectivity: the citizen reimagines their own role as an urban denizen who is doing their part to increase efficiency.

Gordon and Walter critique the ideology, discourse, limited forms of action, and other aspects of the growth of smart (and participatory) cities. They also note that these systems constrain and limit citizen participation even as they maximize the efficiency of existing city systems. However, they don't fully explore the differential impacts of these systems on different kinds of city residents. What happens when we bring race, class, gender, sexual orientation, disability, and/or immigration status into the analysis of these developments? So-called smart city systems have differential impacts on the lived experience and life chances of city residents based on their location within the matrix of domination.

Next, what are the larger structural effects of the spread of design thinking labs and processes across cities? The language of civic innovation is often neoliberal code for the continued shrinkage of the social welfare state. Public programs are converted into design challenges as the first move in the privatization process. Participatory design processes are too often used to generate community-created materials that provide cover for the underlying assumption that the private sector can do everything better, cheaper, and more efficiently. Sustainability is converted from a systems-level analysis that examines the long-term maintenance of a particular process, taking labor, ecology, and social goods and harms into account, into an organizational-level analysis that focuses primarily on the efficiency of a state organization or the potential profitability of a particular firm. The design process itself becomes an exercise in the state feeling good about itself. In the worst cases, participatory design processes are simply used to provide legitimacy for preexisting plans. More typically, a small group of mostly middle-class participants have a chance to suggest minor modifications to processes and plans the guiding principles of which, if not their most significant aspects and detailed clauses, have already been determined according to the interests of incumbent power holders and professional lobbyists.

Overall, deindustrialization and the emergence of empty or abandoned factory zones close to urban cores, coupled with the rise of new social movements, squats, and autonomous social centers, was an important condition for the first wave of hacklabs in the European context. It also seems to be a factor in the Latin American context—for example, in the case of Las Barracas Hacklab in Buenos Aires.[59] A tentative hypothesis might connect hacklabs as explicitly movement-linked, politicized spaces to the moment of local deindustrialization. In other words, hacklabs emerge as factory production shifts locations and abandons middle- or higher-income countries for cheaper labor and more lax environmental regulations in lower-income countries. Later, the urban cores and nearby postindustrial zones are reorganized by an influx of new kinds of capital for globally networked, information-intensive industries, such as software development and biotechnology. Poor and working-class people are pushed out of the urban core as rents soar.[60] Abandoned factories are reclaimed, demolished, and/or refurbished as hip corporate offices or ready-made "live/work" lofts. This largely displaces squatters, artists, political organizers and activists, and social centers. The political hacklabs are either physically displaced or shift gears to accommodate the discourse of neoliberal entrepreneurialism. New spaces are created that are native to this discourse. This narrative isn't generalizable everywhere and is admittedly painted in quite broad strokes. However, understanding these larger patterns might help produce intentional strategies to maintain more hacklabs as movement-linked, rather than corporate-linked, design sites.

Fablabs: Designing Whose Reality?

A fabrication laboratory, or *fablab*, is "a small-scale workshop offering (personal) digital fabrication."[61] Fablabs include tools for design, modeling, prototyping, fabrication, testing, monitoring, and documentation. The idea of fablabs emerged through a collaboration between the Grassroots Invention Group and the Center for Bits and Atoms at the MIT Media Lab.[62] The first fablab was set up in 2002 at Vigyan Ashram in Pune, India; since then, the fablab network has grown to about 1,300 sites (according to the network's website).[63]

Figure 4.2, showing a "typical fablab," is the main, full-page visual of the fablab network's landing page at fablabs.io. It shows eleven people;

Figure 4.2
"A typical fablab." Main image from fablabs.io.

all appear to be male or masculine-presenting, with nine young boys, one young adult, and one older man. All except for one or two appear to be white. Although gender presentation is not the same as gender identity, and a quick reading of this image cannot confirm the race, gender, or other identities of the participants, the image generally conveys the impression of a fablab as a space for white boys to learn about technology together, with intergenerational mentorship and guidance from older men.

MIT professor Neil Gershenfeld, co-creator of the fablab concept, recently coauthored a new book that he positions as a kind of guiding bible for the fablab network: *Designing Reality: How to Survive and Thrive in the Third Digital Revolution.*[64] Copies of the book are freely available to all fablabs—although somewhat ironically, given the basic proposals in the book for the restructuring of global production and consumption systems through local and digital fabrication, the book is not freely available as a downloadable file. The authors include a section about the threat of a "third digital divide" that could potentially extend, and even amplify, inequality based on the differential rollout of digital fabrication technology around the world. However, they conceive of inequality in the broadest terms, as a relationship between what they call *wealthier countries* (what others might describe as nation states that became wealthy through the historical processes of settler colonialism, native genocide, slavery, and/or extractive colonialism and the theft of natural resources) and *poorer countries* (those that were dominated for hundreds of years through military force and occupation under European colonialism). They do mention gender but do not discuss the specific ways that digital inequality is structured by gender, race, class, disability, or migration. They do not consider the intersectional distribution of resources and opportunities, nor does their conception of the digital divide incorporate the concept of the matrix of domination. An intersectional analysis of the distribution of benefits of digital fabrication, as structured by the matrix of domination, would be far more precise than their abstracted, country-level digital divide framework.

In addition, a design justice approach to the question of "designing reality" would involve multiple levels of analysis: "Who will have access to digital fabrication tools?" is an important question, and it is the question that fablabs seek to address. Another might be "What values will guide the use of these tools?" On the one hand, there is clearly a huge opportunity to use digital fabrication to hard-code liberation across a broad range of material objects through the approaches that we discussed in chapter 1. Broader availability of digital design and fabrication increases the opportunity for community-controlled design processes, as we saw in chapter 2, and that include more diverse users and user stories, with more equitable and inclusive distribution of affordances (and disaffordances). However, this opportunity by no means implies that this path is natural, inevitable, or even likely.

Instead, absent an intentional and systematic effort to implement a pedagogy and practice of design justice, fablabs (like hacklabs, makerspaces, and other privileged design sites) too easily become sites for the reproduction of the matrix of domination, despite their promise to radically democratize the means of production of everyday objects. More specifically, fablabs may reproduce patriarchy, white supremacy, and settler colonialism, even as they challenge (if we want to be quite generous) or reconfigure (to be more realistic) capitalism. In many cases, these sites do partially challenge ableism through a common emphasis on assistive technology, such as 3-D printed prosthetics, but typically do so through the individual/medical model of disability, rather than the social/relational model, let alone a disability justice analysis.[65]

The Fab Charter[66] says that the spaces are "open," but it doesn't mention or specify even a desire for diversity and inclusion, doesn't propose a code of conduct of the kind that intersectional feminist spaces frequently employ, and so on. The assumption that making sites "open" makes them inclusive, without specifically addressing race, class, gender, and/or disability dynamics, is common to many privileged design sites. For example, the hacklab design kit, published in 2007 by Hackers on a Plane and credited widely with kicking off the current wave of hacklabs, doesn't discuss race, class, gender, or disability.[67] By contrast, more recently several templates, guides, how-to manuals, and zines promote diverse, inclusive, and explicitly antiracist feminist sites such as events, conferences, and ongoing spaces. For example, consider the approach taken in the *DiscoTech* zine (described in the beginning of this chapter),[68] the Code of Conduct promoted by the ADA initiative to make more gender-inclusive conferences,[69] the AORTA collective facilitation guide for inclusive and antiracist events,[70] and similar documents.

The relationship of fablabs to design justice principles is thus complicated. Fablabs do promote the idea of technological democratization, and they do challenge the idea of technological expertise as an exclusive realm. However, they also participate in the discourse of neoliberal entrepreneurialism. Like hacklabs and makerspaces, fablabs are usually framed as sites where individual subjects can learn STEM skills, better position themselves for technology-related jobs, and create and invent new products and start-ups that can be smoothly (or "disruptively") integrated into global capitalism. For example, one of the most visible outcomes of the digital fabrication movement is MakerBot, a 3-D printing company that appropriated the designs, volunteer community, and momentum around the RepRap open-source printer, patented certain aspects of 3-D printers and 3-D printing software interfaces, then was purchased by multinational company Stratasys for $604 million.[71] The MakerBot 3-D printer can now be purchased at Walmart. As Smith et al. note, "The figure of the design-savvy and networked (social) entrepreneur looms large here."[72]

In some cases, open and collaborative design methods that are the cultural norm in such sites have made their way into business practices: "Rather than seeing openness as a threat, firms are becoming familiar with ways of engaging and appropriating the fruits of collective, alternative, or deviant prototyping and learning how to enclose designs, control marketing and benefit from the diffusion of the resultant products and services."[73] In other words, similar to Tiziana Terranova's critique of free labor for social media platforms,[74] the democratization of production can be seen as a new mode of exploitation where "ideas, design, and research efforts are effectively outsourced to 'free labour' in workshops, but with capital retaining the power to appropriate, enclose and commercialize the most promising fruits of that common endeavor."[75]

Simultaneously, though, "workshops are seen as boosting resilient, cooperative local economic activity based in grassroots initiative, collaboration, control and development. The figure of the community activist has a presence here."[76] Many of the sites discussed in this chapter are also committed to free and open-source software and hardware. As Smith et al. put it, most hacklabs, makerspaces, and fablabs have policies and cultural norms via which "all code, designs, and instructions in the making and repairing of something are made freely available for people to access, adopt and modify, so long as the source is acknowledged and any modifications also become freely available."[77] Many who are active in these sites feel themselves to be participants in what legal scholar and political economist Yochai Benkler defines as *commons-based peer production*, or "decentralized, collaborative, and non-proprietary; based on sharing resources and outputs among widely distributed, loosely connected individuals who cooperate with each other without relying on either market signals or managerial commands."[78] Others identify as tech innovators ushering in an age of personalized manufacturing, mass customization, and a new industrial revolution.[79] Still others feel that hackerspaces, fablabs and makerspaces are crucibles for the formation of technological citizenship.[80] Smith et al. consider them to be key infrastructure for grassroots innovation movements, "not as a new model for transforming production and consumption but, rather, as a real-life laboratory experimenting with grassroots fabrication possibilities in terms of objects, practices, and ideas."[81]

Barcelona's city government is working to support makerspaces, called *Ateneus de Fabricació Digital* (digital fabrication workshops), with a plan to open them in every neighborhood across the city by 2040. These are meant to be key nodes in a production system that will help the city locally manufacture at least half of its needed goods. Similar initiatives have been announced by the municipality of São Paolo and by the Icelandic government. However, critics say the Ateneus plan is "the latest in a series of city makeovers, prioritizing international capital markets and speculative investments in the city over the real needs and aspirations of its residents."[82] Conflict over these city-led makerspaces reached a peak when the city displaced a food bank in one of its poorest districts to set up an Ateneu, leading to an occupation of the site by community activists who demanded the reopening of the food bank and a refocusing of the Ateneu to emphasize job training.

Overall, most hacklabs, makerspaces, and fablabs fail to disrupt the matrix of domination. Some of the key organizers of these sites feel that providing people with digital tools is enough. A few actively resist the idea that these spaces should be part of specific social movements or that they should have an explicitly political program. Instead, they emphasize individual autonomy and the "empowerment" that comes from individuals developing their own ability to hack and make things. Some imagine their spaces as incubators of a future where personal programming and digital fabrication will become ubiquitous at the household level and shared spaces will no longer be necessary—much as shared computers supposedly disappear in advanced economies once "everyone" has a personal computer (this is not really the case, as is evident in any library or school computer lab).

As Smith et al. put it, broader institutional forces are increasingly invested in transforming these sites according to their own agendas, through funding, partnerships, discourse, and more. "If workshops are to genuinely realize [their] transformative potential," they note, they must develop a broader analysis of the role they hope to play in the larger political economy, and become strategic about it, or risk being "pulled into the institutional logics ... that could force design and fabrication activities back onto dominant development pathways."[83] Indeed, corporations, governments, and even military agencies are all extremely interested in these sites. For example, Chevron donated $10 million to the Fab Foundation to establish more fablabs, while the Defense Advanced Research Projects Agency (DARPA) partnered with *Make* magazine.[84] Others, like TechShop (the chain of ten makerspaces that suddenly declared bankruptcy in 2017) have attempted to create a business model out of setting up spaces modeled on these sites, then charging membership fees.[85]

Democratizing access to design tools and skills is truly important. We should laud and support efforts to create spaces where more people can learn how to design, prototype, code, hack, make, and build. At the same time, without intentional intervention, these spaces find it very difficult to fulfill even their own liberal democratic rhetoric, because they end up dominated by white cis men and by middle-class people with free time and disposable income. What's more, if we imagine all such spaces magically transformed into gender-diverse, multiculturally inclusive sites overnight, this would be a huge improvement. However, it would not be enough to realize their truly transformative potential. That requires the communities around such sites to develop their own shared analysis of unjust power (the matrix of domination), how to dismantle it, and the specific role of design, hacking, making, and fabrication in that much larger process. It means development of shared identities beyond the neoliberal entrepreneur. Most of all, it requires the intentional nurturing of deep links between these sites and already existing social movement networks.

Hacking the Hurricane? Hackathons, DiscoTechs, Convergence Spaces, and Other Design Events

So far, this chapter has focused on ongoing design sites like hacklabs, makerspaces, and fablabs. Another key type of design site can be found in temporary events like hackathons and design jams. A hackathon is "an event in which computer programmers and others involved in software development collaborate intensively over a short period of time on software projects."[86] The mythology of hackathons is perhaps best expressed in the 2010 film *The Social Network*. In one scene, a young Mark Zuckerberg presides over what is essentially a frat party, but with computers. Drunken (white, cisgender, male) college student developers gather in a dark basement, bingeing on beer and pizza, competing to solve a coding challenge and thereby win employment at the then-nascent social network site TheFaceBook.com. Many of the dynamics at play in semi-permanent sites like hacklabs also operate, sometimes with condensed intensity, in more temporary or pop-up design sites like hackathons.[87]

Hackathons have become increasingly popular both in the private sector and under the auspices of the neoliberal state. They are understood by corporate managers as potentially effective ways to identify new talent, and therefore as a possible mechanism in the tech sector hiring pipeline. In "Hackathons as Co-optation Ritual: Socializing Workers and Institutionalizing Innovation in the 'New' Economy," sociologists Sharon Zukin and Max Papadantonakis draw from their ethnography of seven New York City hackathons to provide a withering critique: "Hackathons, time-bounded events where participants write computer code and build apps, have become a popular means of socializing tech students and workers to produce 'innovation' despite little promise of material reward ... [Hackathons] reshape unpaid and precarious work as an extraordinary opportunity, a ritual of ecstatic labor, and a collective imaginary for fictional expectations of innovation that benefits all, a powerful strategy for manufacturing workers' consent in the 'new' economy."[88] In short, from a managerial perspective, hackathons provide excellent opportunities for the extraction of free labor. This helps explain the increasing popularity of hackathons within the regular practices of technology firms.[89] As evident in figure 4.3, hackathons have become increasingly popular over the last decade. The state has also adopted hackathons at multiple levels, from city halls to the White House. Symbolically, a government agency running a hackathon signals an embrace of technology, as well as of the solutionist framework of civic technology. Some government actors may cynically organize these events as (primarily) a media spectacle. Others are truly excited by the possibilities of civic tech and its instantiation in the event form of a hackathon.

Many of these same dynamics are at work in the nonprofit sector and in "civic hackathons." On the one hand, nonprofit and civic hackathons are less focused on the creation of profitable new firms; instead, they typically seek to produce social, environmental, or civic innovations. On the other, if anything, this sector is more solutionist than the private sector. The assumption that a "hackathon for good" will be successful if it produces a new app that can help "solve" a social problem runs deep.

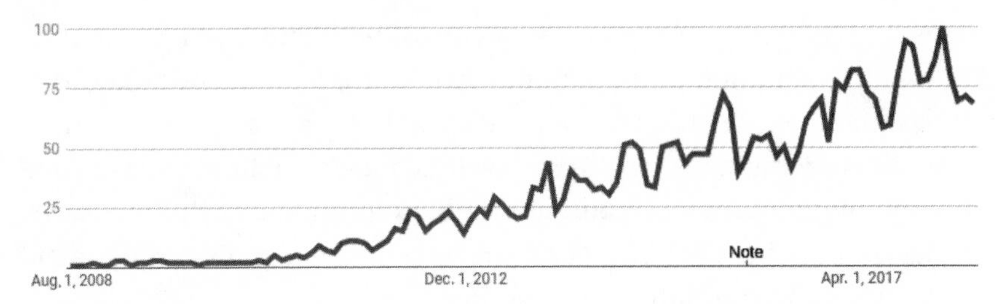

Figure 4.3
Web search interest in *hackathon* over time, from July 17, 2008, to July 17, 2018, from trends.google.com search for "hackathon."

Whether in the corporate, state, or nonprofit sectors, the model of hackathons as prize competitions has also gained prominence in recent years. In this model, teams compete to solve design challenges, typically by producing a prototype web and/or mobile application within a limited time-intensive sprint. Panels of judges review prototypes and pronounce one or more winners. Winners receive prizes, usually consisting of some amount of money, a chance to present the prototype to venture capitalists (VCs), publicity, and perhaps free (as in beer) access to packages of (nonfree, proprietary) software, web services, and tools. Recognizing the potential value (symbolic, material, speculative) of hackathons, a small sector populated by for-profit firms that bill themselves as expert hackathon organizers has emerged. These firms, such as Hackathon.com and BeMyApp, are contracted by other companies to organize and run corporate hackathons.

Hackathons: The Bad

Prompted by his experience with the way media outlets covered a Hurricane Hackers event at the MIT Media Lab, software developer and cooperative economy advocate Charlie DeTar wrote a blog post that powerfully summarizes some of the frequent problems, narrative distortions, and potential benefits of hackathons.[90] DeTar and others have pointed out some of the most frequent problems with hackathons: they're often dominated by white, cisgender men with software-development skills; they tend to be exclusive, normative, and solutionist; they often don't respect the experiential knowledge and tacit expertise of people who deal with the issue area of the hackathon on a regular basis; they nearly always focus on problems and rarely build on existing community assets; and people think hackathons can do things that they usually can't, such as solve big or even little problems, create new products overnight, or 'level the playing field' of innovation through meritocracy. Data journalist and professor Meredith Broussard, in her brilliant book *Artificial Unintelligence*, levies a similar critique of hackathons and the culture around them.[91]

To take another example, writer, artist, and activist Gloria Lin's (2016) ethnography of college hackathons describes how "hackathon spaces cultivate a culture that marginalizes hackers with specific needs, including but not limited to women, people with disabilities, people with non-traditional backgrounds, and even individuals with specific dietary restrictions. By consistently ignoring the health, diet, and care needs of diverse attendees, along with needs based on skill, class, and gender identities, hackathons create an exclusive and hostile environment."[92] Lin describes in detail what this looks like at a UCLA hackathon organized by a company called Major League Hacking. She notes that ideal attendees, organizers, mentors, and judges were all white or Asian cis men. Lin also points out that hackathons often reward misogyny: "At LA Hacks 2014, Wingman, an initial finalist, made headlines as a winning app that analyzed photos of females to determine their promiscuity and whether they'd altered an image to look more attractive. The openness to this type of content at hackathons sends the message that women aren't welcome."[93] In addition, hackathon teams tend to recreate the wheel. The cult of the new and shiny drowns out the quiet call of the well-established,[94] and no one wants to solve the bugs in the old thing that already does what the new thing is supposed to do.

Why does this happen so frequently? For one, under patriarchy, making something "new" is valued more and is better rewarded than caretaking, maintaining, or supporting something "old." Also, starting a new thing is fun; for many, it feels more creative than working on the nuts and bolts of an existing project. There are also context-specific challenges and reward structures at play. If you have limited time to work on a particular problem, it may not make sense to contribute to an existing software project unless it's possible to reach that project's maintainers in real time. Existing systems that address a design challenge are likely to be more complex and feature-robust than a proposed new solution because they have had time to be exposed to real-world use (and real-world challenges to design assumptions). Creating a new big-picture solution in prototype form is generally simpler and carries more immediate rewards than contributing a small improvement to an existing real-world tool. Contributing to an existing project requires contacting and negotiating with the existing developers, maintainers, and community. Creating something new produces attribution, credit, and visibility for its developers, whereas attribution, credit, and visibility for participating in an existing project must, at the very least, be shared.

These problems are widespread across corporate, state, and nonprofit or civic tech hackathons. The solutionist approach to civic hacking is sometimes mitigated by the inclusion of individual(s) from the most affected community on the hack team, but in general the short time span, problem-based (as opposed to asset-based) framing, and product-oriented process of most hackathons makes them a poor fit for deep engagement with the principles of design justice.

In chapter 2, I described the Technology for Social Justice project and the *#MoreThanCode* report (https://morethancode.cc), based on interviews and focus groups with 188 technology practitioners. Many of these practitioners had strong opinions about hackathons. #MoreThanCode participants mentioned many problems with the dominant hackathon model: several noted that most hackathons don't produce working products, that hackathons can bring out weird power dynamics with people competing for leadership, and that women often experience sexism at hackathons.[95] They felt that hackathons often reinforce elite networks and don't usually include the most impacted community members. For example, one noted that most hackathons meant to help low-income people don't usually have the intended end user at the table.[96] As one practitioner who came to tech from legal services put it, "A one day hack for homelessness takes away from the complexity of social justice issues. … You can't just come up with an app and solve the world's problems."[97] In general, study participants said that hackathons often attempt to solve big, underlying problems with technology, when what is needed is democratic consensus, strong social movements, and policy.

Another practitioner, who works at a civic tech unit within Microsoft, noted that there are many new actors, including traditional start-up accelerators, that are increasingly interested in the "tech for good" space. Unfortunately, they said, these new actors mostly seem to be ignorant of the work that has already been done. They provided an example of a civic tech hackathon where they gave a talk: "You can totally create a new call Congress tool, just please know that these other seven ones are there and tell me how yours is different"; however, they said, the civic hackers "ignored my talk ... one of the winners built a call Congress tool."[98]

In short, hackathons are too often sites where the dynamics of structural inequality and unquestioned privilege are reproduced. Like many tech spaces, they tend to be dominated by white, straight, able-bodied cisgender males, masculinist assumptions about technical competence, universalizing discourse, and solutionism. They are too frequently exclusive, alienating to those who don't already feel comfortable in normative tech culture, and dismissive of the difference, experiential knowledge, and domain expertise of marginalized people.

Hackathons: The Good

All that said, hackathons, like other design events, are potentially very valuable sites for the practice of design justice. They are often crucibles of intense and focused learning, making, problem-solving, community building, and play. One #MoreThanCode participant notes that hackathons can be good for connecting domain experts, community members, designers, developers, and researchers.[99] Researchers Robinson and Johnson agree: they argue that city-run hackathons can create valuable spaces for administrative staff to interface with interested publics, provide clear feedback for city administrators about what open data sets are important, "help put open data into public use," and inform future open data releases.[100]

Another #MoreThanCode participant says that hacker camps, such as Chaos Communication Camp and ToorCamp, create comradery; a second points out that hackathons form networks of people who can be mobilized to participate in larger collaborative projects; a third notes that, if well-organized, hackathons can provide an introduction to the use of tech for social justice, as well as pathways to employment.[101] For example, in response to the question "How did you enter this work?," one software developer interviewed for the *#MoreThanCode* report describes a city hackathon, in partnership with Google, as their entry point; another mentions organizing woman-centered hackathons as an important aspect of their own career path.[102]

Many feel that focused pop-ups, hackathons, and hack nights can be valuable if they respond to the real needs of organizations with a lot of domain expertise, instead of focusing on ideas that come from coders.[103] For example, one participant notes that hackathons have been useful in the legal services community to generate ideas for apps to support clients and legal aid workers as they navigate the legal system. A software developer at a worker owned co-op shares that some tech communities, like Drupal for Good, use hackathons to organize pro bono website creation for community-based organizations.[104] According to these participants, there has been a slow, long-term shift toward including community members in design and development processes, including hackathons: "In the old days people used to form teams and rush in and try to fix things, without really even knowing what was broken ... it is no longer just a bunch of programmers in a room. There are now hackathons where actual community members are learning to code and interacting. ... Community members are also teaching programmers about the things they need to sustain and build for the future. That's a really good thing happening."[105]

Rethinking Design Sites through a Design Justice Lens

What are some of the practical implications of design justice for how we organize design sites? Recall that, at its core, design justice is about the fair distribution of design's benefits and burdens; fair and meaningful participation in design decisions; and recognition of community-based design traditions, knowledge, and practices. Some design spaces, increasingly, are oriented toward some or all of these goals.

In their account of hackerspaces (summarized at the beginning of this chapter), Grenzfurthner and Schneider urge the repoliticization of design sites. In part, they imagine this can be accomplished through a rediscovery of theory and history: people in hackerspaces need to learn about where hackerspaces came from, discuss the social developments that they oppose (identify the "anti"), and develop their own theories of resistance and social change. They also call for a shift in the leadership of most hackerspaces, which they note are largely dominated by "benevolent" and informal white, male, nerd elites. They needle US hackerspaces to consider whether they include Black and/or Latinx members, European hackerspaces to be reflexive about whether they have North African or Turkish migrant members, and hackerspaces everywhere to be real about gender balance in membership and leadership: "What is needed is the non-repressive inclusion of all the groups marginalized by a bourgeois society just as it had been the intention of the first hackerspaces in countercultural history. If we accept the Marxian idea that the very nature of politics is always in the interest of those acting, hackerspace politics are for now in the interest of white middle-class males. This needs to change."[106] Beyond representational politics, design justice compels us to develop a range of strategies to explicitly relink design sites to social movement organizations that are rooted in marginalized communities.

Happily, many people are already working to create new, radically inclusive design sites, to transform existing sites, and to explicitly relink hacklabs, hackerspaces, and hackathons to social movements. Some spaces embrace the hacker ethic while striving to be radically inclusive; examples include Design Studio for Social Intervention in the Boston area; Intelligent Mischief in Brooklyn; LOLspace and Double Union in the Bay Area; the Afro-Brazilian hackerspace Ujamaa in Salvador de Bahia, Brazil; and so many more.

Activist and researcher Sophie Toupin provides a brilliant and definitive account of the rise of intersectional feminist hackerspaces in the United States. She begins by summarizing intersectional feminism, then reviews what hackerspaces are and provides a brief history of their spread from Europe to the United States. Next, Toupin traces the origin of feminist hackerspaces in the United States to the Geek Feminism Wiki, the Ada Initiative, and a presentation by Seattle Attic on "how to build a feminist hackerspace" at the third Ada Camp in San Francisco in 2013. She shuttles back and forth between concise description of theoretical developments in intersectional feminist thought, a history of hacker, geek, and maker feminists, and discussion of the recent establishment of intersectional feminist hackerspaces like Mz* Baltazar's

Laboratory in Vienna, Liberating Ourselves Locally in Oakland, Mothership Hackermoms in Berkley, Seattle Attic, Flux in Portland, Double Union in San Francisco, and Hacker Gals in Michigan.[107] Liz Henry, in a widely circulated 2014 article for *Model View Culture*, described the creation of the Oakland-based feminist hackerspace Double Union and located it alongside the history of feminist organizing within technology.[108] Another makerspace that positions itself as antiracist, feminist, and activist-oriented is the Sugar Shack in Los Angeles, open since 2001.[109]

Some makerspace and fablab organizers have explicitly incorporated a liberatory political vision. For example, De War's fablab in Amsterdam appropriated the MIT model, opened a grassroots fablab without permission, and now focuses on hacking production, consumption, and the broader economy through the lens of sustainability and resilience.[110] This has been described as a *grassroots insurgency* and appropriation of the fablab model.[111] There are also many individuals within the fablab network who seek to transform fablabs into more inclusive spaces.

There are also many important parallel processes that aren't necessarily called *hacklabs* or *hackerspaces*. For example, in the United States, under the Broadband Technology Opportunity Program (BTOP) section of the American Recovery and Reinvestment Act (ARRA, more commonly known as the Obama Stimulus Bill), organizers with the Media Mobilizing Project in Philadelphia were able to leverage federal funds to resource community computer labs that combined broadband access with political education and media production workshops.[112] Community-based workshops where people learn design skills, share access to tools, explore hacking and repair culture, and generally challenge the disposable logic of consumer culture are increasingly widespread. As Smith et al. put it: "To hack open a device designed for obsolescence, and to repair it and upgrade it and then to share freely that knowledge about the device and its workings is a deviant act within the logics of cognitive capitalism. ... The question is whether these initiatives ... can connect to movements that are seeking pathways organized to alternative logics of sustainability and social justice."[113]

Although they are beyond the scope of this chapter, libraries also have long been important to the democratization of access to knowledge, and there's a growing trend to develop libraries as sites where people can learn about and explore digital design.[114] Another interesting site is Computer Clubhouses, where educators like Jaleesa Trapp work with low-income youth, girls, and/or gender-non-conforming kids of color to build their design skills and their support networks.[115] Schools are, of course, also crucial sites for the development of a praxis of design justice. I further discuss pedagogies in chapter 5.

Hacking Hackathons: Models for More Inclusive Design Events

Just as some more permanent design spaces can potentially support the goals of design justice, so can short-term design events. Active social movements have already developed multiple models for hackathons and other design events that are more closely aligned with design justice principles. For example, during the height of the global justice movement in the late 1990s and early 2000s, before there were hackathons and DiscoTechs, mass mobilizations and convergences often featured media/tech labs. As anthropologist Jeffrey Juris describes, these temporary labs were embedded within larger global justice convergence spaces where all kinds of movement activity were coordinated, so they served as tech and media organs within a larger social body. They were sites of sociotechnical innovation, knowledge exchange, and community building, but they didn't exist in a vacuum. They responded most directly to the particular project at hand, which was to effectively organize communications, connectivity, and ICT infrastructure to support large-scale mobilizations and independent media coverage of those mobilizations.[116] I personally participated in and helped organize these kinds of spaces at protests against the World Trade Organization (WTO) ministerial meeting in Cancun (Hurakan Cancun), at the Free Trade Area of the Americas (FTAA) IMC that was organized during protests in Miami in 2003; the We Seize! Hub at the World Summit on the Information Society (WSIS) in Geneva in 2003;[117] the Polimidia Lab at the OurMedia conference in 2004;[118] and the Twin Cities Indymedia Center that was organized during protests against the Republican National Convention in 2008. In 2012, similar media and tech spaces were organized at many of the occupy encampments. There were media tents at Occupy Boston, Occupy Wall Street, Occupy London, and many more. At Occupy DC, there was a tech space and a working group that, among other projects, organized a mesh wireless network for the camp and prototyped portable battery mounts for small computers to enable consistent, high-quality livestreams from the camp even during marches and in the event of dislocation by the police.[119]

Today, there is a growing community of people and organizations that works to create and share models for inclusive hackathons or for hackathon-like events that capture the spirit, energy, and positive possibilities of hackathons while transforming their too-often exclusionary tendencies. This includes groups like Geeks Without Bounds, Aspiration Tech, the Detroit Digital Justice Coalition, and EquityXDesign, among many others. We have already discussed DiscoTechs and the context of the Allied Media Conference from which they emerged. Other conferences and events that emphasize diversity and inclusion include MozFest, the Internet Freedom Festival, CryptoHarlem meetups in New York City, CryptoParties in Brazil, Encuentros Hackfeministas throughout Latin America, and the Tech Lady Hackathons in DC. The Lesbians Who Tech Summit[120] provides a physical meetup and networking space for lesbians working at all levels of technology industries. Trans*H4CK is a series of hackathons by and for trans* and gender-non-conforming people, with local events in San Francisco, Boston, and other cities.[121] In Latin America, there are International Development Design Summits (*Cumbres Internacionales de Diseno para el Desarollo* [IDDS]). WhoseKnowledge.org is "a global campaign to center the knowledge of marginalized communities (the majority of the world) on the internet ... we work particularly with women, people of color, LGBTQI communities, indigenous peoples and others from the global South to build and represent more of all of our own knowledge online."[122] The campaign organizes resources, how-to guides, summits, and hackathon-like *knowledge sprints* where participants edit Wikipedia together to recenter marginalized people, histories, and knowledge.

The accessibility technical community organizes inclusive design events under the moniker #A11yCAN Hackathons. These focus on "the design of products, devices, services, or environments that can be used by people with disabilities."[123] The Make the Breast Pump Not Suck Hackathon and Policy Summit focused on improving both the design of breast pumps and the policies and norms that push breastfeeding people, especially low-income PoC, away from breastfeeding. The series organizers worked hard to create a space that was inclusive, centered the experience and expertise of low-income women of color and reproductive justice organizations, and supported participation by mothers with infants and young children.[124]

Gloria Lin has also noted the emergence of more inclusive hackathons, such as "Technica, an all-women hackathon at the University of Maryland, College Park ... [which] incorporates yoga breaks in its schedule, which allows hackers to practice self-care. Hack Davis at the University of California, Davis brands itself as a 24-hour social hackathon."[125] As Lin describes, "Such hackathons push hackers to reflect on why they are doing the work they do, push for the ideas and welfare of marginalized communities in the tech sphere, and do so on the terms of their wellbeing and safety."[126]

In a 2015 conference paper for SIGCSE, education technology scholar Gabriela T. Richard and her coauthors describe StitchFest, a hardware hackathon focused on using LilyPad Arduinos to design wearables under a theme of "wear and care."[127] They argue that they were able to increase women's participation rate through targeted recruitment, a thematic focus, and offering participants particular kinds of materials. They also summarize the recommendations from the National Center for Women and Information Technology (NCWIT) about how to recruit more women and minorities to coding competitions: "(1) including promotional materials that feature females and a range of students, (2) actively recruit females, (3) provide ongoing encouragement, (4) allow participants to create projects that appeal to them, (5) encourage mixed teams with experienced and inexperienced members, (6) host a tutorial or how-to event, (7) focus on learning and different ways to win, (8) include female mentors, educators and judges, (9) make sure the space is accessible to all, and (10) educate others involved."[128]

Along similar lines, #MoreThanCode participants had many specific suggestions for how to effectively organize design sites to be more diverse, inclusive, and useful. One pointed out that the way you organize a hackathon greatly influences who will show up, and that aspects such as the day of the week, time of day, location, and highlighted speakers are all important.[129] Another described working with the NYC Parks Department to hack on thirty years of open tree data; this hackathon included volunteer tree stewards, neighborhood association staff, city-wide nonprofits, and parks department staff, as well as people from the tech, data, and design communities.[130] One practitioner described the process of creating intersectional, feminist, PoC-led makerspace in Oakland; another shared their experience of setting up a city innovation lab in New York City.[131] Several shared that social justice tech organizations, especially May First/People Link, have long organized tech convergences that gather activists to identify the ways that technology might be used to most effectively build social movement power. In the end, the *#MoreThanCode* report provides the following key recommendations for creating more diverse and inclusive technology design sites: gather and publicly share diversity data; set public, time-bound diversity targets; and "transform conferences, convenings, meetups, and other gatherings to be far more diverse, inclusive, accessible, and affordable. Adopt best practices for inclusive events, such as the DiscoTech model. Do the same at key sites such as libraries, universities, community colleges, hacklabs, and makerspaces."[132]

In general, it should be possible to organize more hackathons according to these recommendations, as well as in accordance with other principles of design justice: most crucially, to include those affected by the domain area of the hackathon on the organizing committee. It is also crucial to encourage or require teams to include people with lived experience of the event domain, and to provide support and scaffolding to make this possible. Other common best practices include the following: develop clear codes of conduct; provide financial support to enable more inclusive participation; and create community advisory boards. As noted in chapter 3, it is also important to develop framing language together with community-based organizations that work on the issue; scope design challenges in consultation with CBOs; and consider asset-based framing rather than default to problem framing. In addition, directly affected people can be mentors for design teams; as well as judges to provide critique and feedback to project teams (whether or not there will be a "winner"). Publicity about design events can highlight the work of existing CBOs, alongside event organizers.

Many #MoreThanCode interviewees noted that at design events, organizers should pay attention to participants as whole human beings. For example, this means that it is important to consider food, bio breaks, accessible bathrooms that are friendly to all body types and genders, comfortable spaces to nap or relax, and decent lighting; provide childcare so that parents can participate; provide a clean, comfortable space to nurse or to pump breast milk; and consider holding events in venues that are familiar to community members. If there are good reasons to hold the event in another kind of venue, organize transportation and other logistics support so that community members can more easily attend; choose locations that are friendly to more than just the "usual suspects," and consider transportation, food, child care, translation, and accessibility.

Most recently, in 2019, students in my Codesign Studio course at MIT focused on working with community-based organizations to "hack hackathons," in other words, to support radically inclusive and accessible design events. As her final project, doctoral student, designer, and software developer Victoria Palacios conducted a literature review of existing suggestions for how to organize better design events, reviewed the lessons that emerged over the course of the semester, and synthesized them all into a set of extremely useful guidelines, freely available at bit.ly/designeventguidelines.

Ultimately, if the master's tools can never be used to dismantle the master's house, as Black lesbian feminist writer, poet, and activist Audre Lorde stated so powerfully,[133] can hackerspaces, makerspaces, fablabs, and hackathons be sites where we develop new kinds of tools? Perhaps, or perhaps not, but either way new tools won't matter if we use them to follow the master's architectural plans. By following design justice principles, design sites might be transformed into feminist, antiracist spaces that are not only truly inclusive, but also organized to explicitly challenge, rather than tacitly reproduce, oppressive systems. Furthermore, they can link with social movements led by those who are multiply burdened under the matrix of domination, in order to help develop plans for new kinds of dwellings.

5 Design Pedagogies: "There's Something Wrong with This System!"

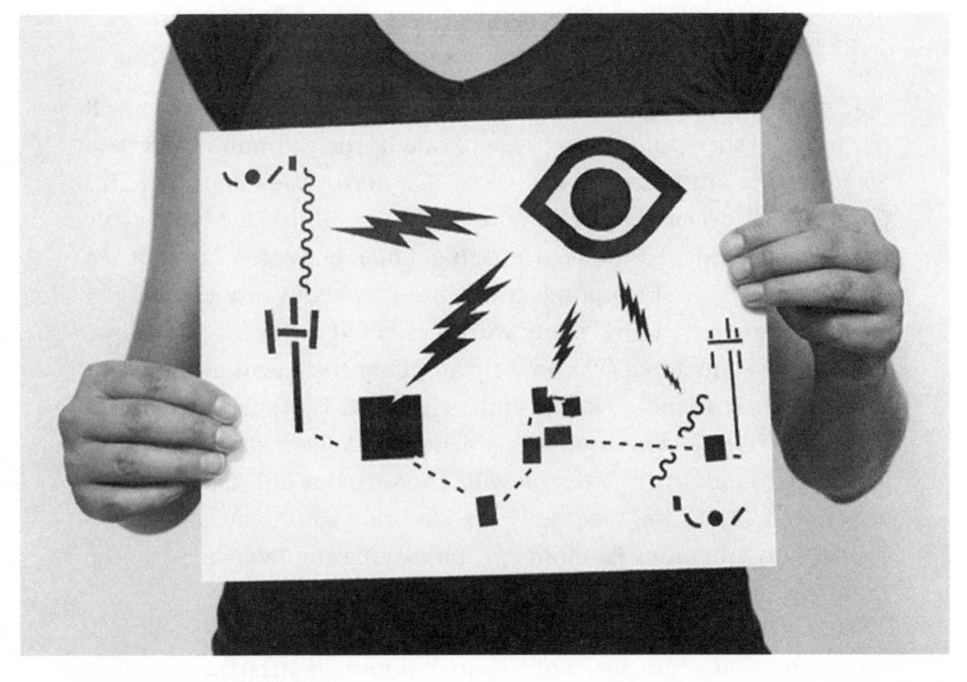

Figure 5.1
Surveillance. Collage by the author, used as the poster image for the 2013 MIT Codesign Studio. Image produced at the Name That Tech workshop by the Work Department at the Allied Media Conference in 2011. Photo by Nina Bianchi.

> Critical pedagogy seeks to transform consciousness, to provide students with ways of knowing that enable them to know themselves better and live in the world more fully.
>
> —bell hooks, *Teaching to Transgress*

> I insist that the object of all true education is not to make [people] carpenters, it is to make carpenters [people].
>
> —W. E. B. Du Bois, *The Talented Tenth*

> Oppressed people, whatever their level of formal education, have the ability to understand and interpret the world around them, to see the world for what it is, and move to transform it.
>
> —Ella Baker

"Are you ready to fight?" The question rings out across the room. It resonates in the strong, clear voice of one of the community organizers from City Life/Vida Urbana (CL/VU), a Boston-area housing rights group. "Yes!" declares a Black woman in her mid-thirties, holding City Life's signature symbolic sword and shield high above her head. "Then we'll fight with you!" responds the roomful of about fifty community members, some standing, many with fists raised in the air. Most have themselves come to CL/VU for help stopping their own evictions, or those of their friends, family, and neighbors. The shield represents the power of collective community action to defend against evictions and foreclosures; the sword represents the weapons of legal and media action that CL/VU has repeatedly used to bring banks, predatory lenders, and unscrupulous landlords to the negotiating table and to keep hundreds of families in their homes.

This is no easy task, in the context of ascendant neoliberal federal, state, and municipal policies that prioritize attracting business and investors to urban cores, and that "sanitize" cities to make them "safe" for young (mostly white) professionals to live in after decades of disinvestment and white flight.[1] Indeed, since the mortgage crisis and financial collapse of 2008, Black families in particular have been particularly hard-hit: Black people lost over 240,000 homes in the crisis;[2] Black homeownership rates are still 6.6 percentage points below their mid-2000s peak; and the majority (51 percent) of Black families still live in high-poverty areas.[3] Housing rights organizing is especially important in this context. Since 1973, CL/VU has successfully organized in the Boston area to create over four hundred units of affordable housing, form tenant associations in over forty buildings, and keep more than eight hundred families in their homes, through comprehensive strategies of eviction defense, legal action, and group renegotiation with lenders, banks, and landlords.[4] The organization is a cofounding anchor of the national Right to the City Coalition, a network of community-based organizations (CBOs) that works across the United States to fight the displacement of low-income people, the majority Black, Indigenous, and/or people of color (B/I/PoC), from their historic urban neighborhoods.[5]

I am there with a group of twenty MIT students, both undergraduate and graduate, from the Civic Media Collaborative Design Studio, a course I have taught at MIT since 2012. In the course catalog (https://codesign.mit.edu), the Codesign Studio (as it is known) is described as follows:

> We will be working to design and develop real world projects, grounded in the needs of CBOs. As a student in the course, you will be part of a co-design team led by a partner organization. … The studio is also a space for shared inquiry into the theory, history, best practices, and critiques of various approaches to community inclusion in iterative stages of project ideation, design, implementation, testing, and evaluation. The Civic Media Codesign Studio approaches communities not as (solely) consumers, test subjects, "test beds," or objects of study, but instead imagines them as co-designers and coauthors of shared knowledge, technologies, narratives, and social practices. Our goal is twofold: to develop an understanding of the ways that technology design processes often replicate existing power inequalities, while at the same time, moving beyond critique towards community coauthorship, as much as possible within the constraints of any given project.

This semester, one of the student teams has partnered with CL/VU to design a media project that will meet an organizational need while connecting to the skills and interests of the students. They work closely with Mike Leyba, communications director for CL/VU. Together with Mike, and with feedback from CL/VU members, they have decided to develop a set of modified carnival games that illustrate inequality in Boston-area housing markets. Their project, titled Change the Game, remixes popular carnival games like whack-a-mole, cornhole, and the shell game. These modified games are designed to be played in public places where they will attract attention, help educate people about

Figure 5.2
Cover art for the Change the Game toolkit, by Ed Cabrera and Triana Kazaleh Sirdenis, for City Life/Vida Urbana.

housing rights, and engage new potential members and allies. The design team has produced a pamphlet about the games to hand out to players and spectators; the pamphlets contain key facts and information about housing, gentrification and displacement, and community organizing for the right to the city in Boston. The games are also designed to shift the narrative around the housing crisis, challenge the idea that the crisis is over, and highlight CL/VU's three ongoing campaigns: fight eviction and foreclosures, resist gentrification, and shed light on real estate investors who turn a profit on foreclosed homes.[6]

After the CL/VU organizing meeting, our class gathers to debrief, discuss what we've learned, and share our feelings. One student, a sophomore studying computer science, describes a moment that they found particularly moving: when the new CL/VU member narrated her personal story, she described how she and her young daughter were evicted from their apartment, ran out of family and friends' couches to crash on, and spent several months sleeping in her car. The MIT sophomore, with a slightly bewildered, slightly angry tone, says: 'There just must be something wrong with a system that would make that woman and her child have to sleep in her car!' When she says "system," from previous conversations I and the whole class understand that she is thinking quite literally in terms of systems design. She wants to find the design errors that result in such a blatant injustice.

This experience provided a key learning moment for participants in the Codesign Studio. I recount it here to ground this chapter's call for *pedagogies of design justice* in concrete experience, with all its attendant messiness, rather than in abstract theories about education. Fundamentally, this chapter asks us to reflect on the following question: "How might we teach and learn design justice?" I don't believe there is only one way to answer this question, which is why I use "pedagogies" in the plural form.

Popular Education: Foundation for Design Justice Pedagogies

That said, I do believe that pedagogies of design justice must be based firmly upon the broader approach known as *popular education* (*educación popular*, in Spanish), or *pop ed*, as it is often called by practitioners in the United States. Pop ed was originally developed by the radical Brazilian educator and philosopher Paulo Freire. During the years of military rule in Brazil, Freire was a political prisoner and then lived in exile; after democratization, he returned to become the secretary of education for the city of São Paulo. In his widely influential book *Pedagogy of the Oppressed*, Freire denounces what he calls the *banking model* of education, in which an educator, positioned as the expert, attempts to deposit knowledge in the mind of their students. Instead, he encourages critical pedagogy, where the role of the educator is to pose problems, create spaces for the collective development of critical consciousness, help to develop plans for action to make the world a better place, and develop a sense of agency among learners.

Freire focuses on developing critical thought together with action, in a cycle he refers to as *praxis*, a Greek term originally referring to "practical knowledge for action."[7] Freire defines it as "reflection and action upon the world in order to transform it."[8] In other words, for Freire and for popular educators inspired by his work, the goal of education is to transform oppressed individuals into subjects who engage in collective action to transform their conditions of oppression. In Brazil and across

Latin America, popular educators, many using Freirian methods of critical pedagogy and praxis, taught millions of rural peasants and urban poor people how to read and write while working together to develop a collective analysis of political oppression and to organize powerful social movements that helped end military dictatorships across the region.[9]

Pop ed has also long played an important role in US social movements, especially the Civil Rights movement. For example, the Highlander Research and Education Center, founded in 1932 by educator Myles Horton, is a social justice leadership training school and cultural center that has used pop ed for decades to build grassroots leadership within movements for civil rights, organized labor, and environmental justice, among others. At Highlander, Horton taught and worked with Civil Rights luminaries including Martin Luther King, Jr., Rosa Parks, and John Lewis. Highlander articulates the key principles of pop ed as follows:

1. Education is never neutral: it either maintains the current system of domination, or it is designed to liberate people;
2. Relevance: Pop Ed engages with issues that people care deeply about;
3. Problem-posing: all participants have the capacity to think, question, and act, and Pop Ed is about identifying the root causes of problems that people want to change;
4. Dialogue: no one knows everything, but together we know a lot, if we listen to each other;
5. Praxis: real learning takes places through the cycle of reflection and action to transform the world;
6. Transformation: Pop Ed is about engaging communities to transform individuals, communities, the environment, and the broader society.[10]

I believe that rethinking design education so that it is underpinned by these principles will be crucial to the larger project of design justice. Happily, this is already beginning to take place.

Pop Ed Takes on Technology and Design

In the spring of 2017, a gathering of social movement technologists (mostly PoC, half women and femmes, and many queer and trans* identified) met at the Highlander Center to strategize about how to use technology for liberation. One of the outcomes was a joint statement that included the following: "Currently technology is being developed, controlled, and owned by the ruling class and used in their interests to maintain a brutal system of superexploitation and oppression. We want a shift in the underlying logic of how technology is created and used. Instead of being used as a tool to divide and conquer, we believe technology must be taken back by the people and used as a tool of liberation."[11] Coauthors and signatories include movement technology organizations like May First/People Link, Progressive Technology Project, Aspiration Tech, Palante Technology Cooperative, and the Detroit Community Technology Project, as well as groups that work at the intersection of technology and other areas, like 18MR.org, Equality Labs, Data for Black Lives, the Stop LAPD Spying Coalition, the Center for Media Justice, and Project South. How are these (and other) organizations putting pop ed principles into practice in the design of new technologies?

Project South, based in Atlanta, Georgia, is a social movement organization that, like Highlander, has used pop ed to build community power for decades.[12] Since 1986, Project South has used pop ed to organize people in the struggle against poverty, violence, and racial injustice. Recently, it has also developed a focus on community control of communications, media, design, and technology. Beginning in 2015, it worked with Global Action Project, Research Action Design, and the Transformative Media Organizing Project to design and develop the Movement History Timeline Tool, an interactive timeline generator for documenting social movement history and for facilitating workshops that link people's personal struggles to historical developments.[13] Project South has also helped gather social movement technologists to build shared analysis and goals at convenings such as the Highlander meeting, the United States Social Forum, and the Allied Media Conference.

Pop ed has also been influential in West Coast community organizing histories. In Los Angeles, the Institute of Popular Education of Southern California (IDEPSCA) is a community-based organization with over thirty years of experience organizing immigrant communities through pop ed methods. IDEPSCA, whose motto is "reading reality to write history," is an anchor member of both the National Day Labor Organizing Network (NDLON) and the National Domestic Workers Alliance (NDWA), two nationwide networks that have managed to advance labor rights and quality of life for some of the most marginalized people in the United States. IDEPSCA and both of these national networks have applied pop ed approaches to technology design for years. For example, while I was a graduate student in Los Angeles, I worked with IDEPSCA for five years on VozMob, a popular education and participatory design project focused on appropriating mobile phones to amplify the voices of immigrant workers.[14] Most recently, NDWA used participatory design methods with home cleaning workers to develop and launch Alia, a platform for portable benefits that is enabling some of these workers to access health insurance and paid time off for the first time in their lives.[15]

A pop ed approach to technology design also includes efforts to develop shared understanding of ICT infrastructure. For example, the Center for Urban Pedagogy in New York City worked with VozMob, the Media Mobilizing Project, People's Production House, and the Center for Media Justice to develop *Dialed In: A Cell Phone Literacy Toolkit*: a set of pop ed materials and design workshops to help people learn about how mobile technology works. *Dialed In* includes units about cellular towers, multimedia messaging, and gateways between the mobile telephony system and the internet. It also includes learning modules about mobile surveillance technologies and examples of how social movements use mobile technology for emancipatory ends.[16]

The Detroit Community Technology Project (DCTP), guided until very recently by community technologist, educator, and artist Diana Nucera (also known as Mother Cyborg) uses pop ed methods to build community capacity to understand, design, build, and maintain wireless internet infrastructure and other technologies. DCTP was founded in 2012 as a project of Allied Media Projects and the Open Technology Institute (OTI). DCTP's digital stewards, themselves community residents, work with other residents, local businesses, and anchor institutions to design, install, and maintain wireless mesh networks and to develop policies to govern those networks. By 2018, DCTP's work had grown to five Detroit neighborhoods, seven New York City locations (starting in Red Hook, Brooklyn), and eleven sites around the world. The organization had secured federal funding to scale up its work through the Equitable Internet Initiative. Its pedagogy, and many of the pop ed workshops that it has developed, is extensively documented in the *Teaching Community Technology Handbook*, a series of zines, and other educational materials that are available on the project website.[17]

These organizations, and many others like them across the United States and around the world, already use pop ed to organize their communities and to engage everyday people in the design and development of new technologies. They have been influential not only in the social struggles of previous generations, but also in the new wave of intersectional social movements that are building power today and that are poised to reshape our world for generations to come.

Multiple Liberatory Frameworks for Design Pedagogies

In addition to those who explicitly use a pop ed framework for teaching and learning about design, there are many other pedagogical approaches that use different terms but are closely aligned with pop ed principles. These include critical community technology pedagogy, participatory action design, data feminism, and certain aspects of constructionism, as well as some strands within digital media literacy.[18] The growing call to decolonize design pedagogy is also aligned with the design justice principles. Although I don't have space here to explore each of these approaches in depth, hopefully the following brief summaries will help locate design justice pedagogies within a broader and rapidly developing constellation of allied efforts.

Designer, educator, and former MIT Codesign Studio participant Maya Wagoner developed the concept of *critical community technology pedagogy*, an approach that "demystifies systemic power inequalities, involves a multi-directional learning process, results in transferable skills, and constructs a new world as it constructs knowledge."[19] In her masters' thesis at MIT, Wagoner posits critical pedagogy as fundamental to the ongoing development of liberatory design practice. She describes examples of this approach in the real world and develops case studies of the Civic Lab for Environmental Action Research, the Detroit Digital Justice Coalition Data DiscoTechs, and the Center for Urban Pedagogy's Urban Investigations.[20]

Alongside the many community-based pop ed design projects mentioned earlier and those analyzed by Wagoner, there is an extensive history of self-organized design workshops, schools, and mutually supportive spaces for learning graphic design entirely outside of educational institutions. In a blog post for the Walker Art Center titled "Never Not Learning," João Doria documents recent workshops of this nature, including A Escola Livre and Escola Aberta (Brazil), Asterisk Summer School (Estonia), Maybe a School, Maybe a Park (California), the Registration School (United Kingdom), the Van Eyck Summer Design Academy: Digital Campfire Series (the Netherlands), and the Parallel School (instances have been organized in Brno, São Paolo, Cali, Leipzig, Lausanne, and London, among other places).[21] Design justice is an approach that already finds resonance in many of these workshops, but its principles might help strengthen them and make them more intentional about the communities that they include.

Another kindred framework is participatory action design (PAD). In a 2007 article about the PAD method, scholars Ding, Cooper, and Pearlman trace its roots to participatory design in the 1970s, led by software developers who worked together with the Iron and Metal Workers Union in Norway (as described in chapter 2). The authors discuss how the University of Pittsburgh's Quality of Life Technology Engineering Research Center[22] uses PAD to develop systems to enhance the quality of life of people with disabilities. The approach brings together people with disabilities (the term used by the authors), engineers, social scientists, family members, and caregivers, and they emphasize the inclusion of end users throughout the product development process.[23] They also describe their efforts to teach PAD through a ten-week intensive program where engineering students interact with end users in product and systems design, as well as a course in Quality of Life Technology Ethnography that provides students with an opportunity to break design out of the lab early on in their careers.[24]

In their 2019 book *Data Feminism*, data scientists, artists, researchers, and educators Catherine D'Ignazio and Lauren Klein include a section titled "Teach Data Like an Intersectional Feminist."[25] They describe how current approaches to teaching data science reproduce oppression when they model a world where elite men lead; data science is abstract and technical; there is little (if any) room for considering the ethics and values of data collection, cleaning, and use; and the learning goal is individual mastery of concepts and technical skills. In contrast, they propose an intersectional feminist approach to the pedagogy of data science, grounded in values of equity and co-liberation. The authors present a compelling argument, grounded in real-world classroom examples. They outline key elements for a feminist pedagogy of data science as follows:

1. Listen to and engage with those most affected by a problem, as in the work of EquityXDesign, anti-oppressive design, and the Design Justice Network;[26]
2. Teach data science in a way that honors context, respects situated knowledge, and makes it clear that data is never "raw," as in the work of Data Basic, data biographies, and data user guides;[27]
3. Emphasize the use of data to create shared meaning over individual mastery, as in the Detroit DiscoTechs or the Data Culture Project;[28]
4. Address, rather than mask, the politics of what gets counted and what does not, as in OpenStreetMap, the Public Laboratory for Open Technology and Science, or Princesa Bathory's ongoing mapping of femicides, Gwendolyn Warren's map of children killed by white commuters in Detroit, BlackLivesMatter's map of police violence, Mimi Onuoha's list of Missing Data Sets, or ProPublica and NPR's crowdsourced reporting on maternal mortality;[29]
5. Teach data science that values ethics, emotions, and reason, not only reason, as in Tahir Hemphill's Rap Research Lab or Rahul Bhargava's Data Murals.[30]

These principles are closely aligned with the principles of popular education, as articulated by the Highlander Center (discussed earlier), as well as with the Design Justice Network Principles, as described in this book's introduction.

Constructionism Another strand of design pedagogy that has been somewhat influential in computing pedagogy is constructionism. Although not explicit about race, class, gender, or disability politics, this is a pedagogical approach that centers context, situated knowledge, and learning by doing. Constructionist learning theory and pedagogy was developed by Seymour Papert, a mathematician, computer scientist, and educator who contributed to the development of artificial intelligence and was one of the creators of the Logo programming language for children. Papert built atop child development psychologist Jean Piaget's work.[31] Piaget rejected the idea that learning takes place when an educator transmits a piece of information to the learner's brain—in Freirian terms, the *banking method* of education. Instead, for Piaget, learning is experiential: it takes place through an active process where the learner develops the ability to modify or transform an object or idea. Papert took Piaget's theories and synthesized them into constructionism's two central concepts: first, that learning is a reconstruction, rather than a transmission, of knowledge; second, that "learning is most effective when part of an activity the learner experiences as constructing a meaningful product."[32] Based on these ideas, Papert helped create Logo, LEGO Mindstorms, and the (problematic) One Laptop per Child (OLPC) project.

In constructionist pedagogy, similar to pop ed, teachers act as facilitators to help students achieve their own learning goals using problem-based learning.[33] Problem-based learning works best when problems are part of larger, ideally real-world tasks; learners are supported to take ownership of the problem; the task is appropriate to the learner's level of understanding and ability; the learner must reflect on what is being learned and how they learned it; and the educator encourages the learner to test their ideas in various contexts.[34]

Mitchel Resnick, a professor at the MIT Media Lab's Lifelong Kindergarten (LLK) group who studied with Papert, continues to develop these ideas and has applied them to the creation of several widely used pedagogical tools. Resnick's work includes, among other things, key contributions to the development of LEGO Mindstorms, meant to teach the principles of robotics, and Scratch. Scratch is a programming language designed for kids, as well as a growing community of thousands of young people who use the software to create interactive projects. Scratch was designed according to constructionist principles to be a language with a low floor (easy for new entrants), wide walls (supports many kinds of projects), and high ceiling (more advanced users can create very complex projects). Ultimately, for Resnick and other creators of Scratch, "there needs to be a shift in how people think about programming, and about computers in general. We need to expand the notion of 'digital fluency' to include designing and creating, not just browsing and interacting."[35]

Resnick and many of his students at LLK are also deeply concerned with persistent educational inequality that disadvantages girls, low-income youth, and/or youth of color and blocks the democratization of computing skills and knowledge. For example, in "The Computer Clubhouse: Technological Fluency in the Inner City," Resnick et al. narrate the history of the Computer Clubhouse, a joint effort between the MIT Media Lab and the Computer Museum to bring computing and software literacy to Boston-area youth who might otherwise not have access to computers. The authors note that many efforts to address digital access inequality focus on providing computers to schools and on teaching children basic computing skills such as word processing. Instead, Resnick et al. argue that the goal should be technological fluency, or young people's ability to fully incorporate computers and digital technology into their own creative practices. They describe how in the 1990s, young people at the Computer Clubhouse learned how to digitally photograph their artwork, import photos into the computer, manipulate the images with software, and design and print out comic books. From this experience, Resnick describes four principles for technology educators: support learning through design experiences, help youth build upon their own interests, cultivate "emergent community," and create an environment of respect and trust.[36] Resnick also maintains that design activities are crucial to learners' experience in the Computer Clubhouse. Design activities encourage creative problem solving, nonbinary (as opposed to right vs. wrong) thinking, problem and solution ownership by the designer, a sense of audience, and a context for reflection and discussion.[37]

Resnick summarizes the core of constructionism in the following two principles: first, "people do not get ideas, they make them." Second, "people construct new knowledge with particular effectiveness when they are engaged in constructing personally meaningful products."[38] Accordingly, in a constructionist pedagogy of design justice, learners should make knowledge about design justice for themselves and do so through working on meaningful projects. Ideally, these should be developed together with, rather than for, communities that are too often excluded from design processes.

Decolonizing Design Pedagogies Along with the shifts in design pedagogy toward community-led processes, intersectional feminist principles, and learning by doing described here so far, the idea of decolonizing design pedagogy is gaining steam. Decolonizing design involves decentering Western approaches to design pedagogy, while centering design approaches, histories, theories, and practices rooted in indigenous communities. For example, Dori Tunstall, the new dean of the Design School at OCAD Toronto, is explicitly working to decolonize the design school curriculum.[39] Sadie Red Wing, a Lakota/Dakota graphic designer best known for her work designing visual materials for the *Mni Wiconi*/Water Is Life struggle at Standing Rock, teaches a course on decolonizing design at the University of Redlands. Others currently focused on decolonizing design pedagogy (in the North American context) include Pouya Jahanshahi at Oklahoma State University, Kali Nikitas at Otis College of Art and Design, Ian Lynham at Vermont College of Fine Arts, Steven McCarthy at the University of Minnesota, and Elizabeth Chin at the ArtCenter College of Design. Designers, scholars, and activists involved in this approach are gathering resources at the site decolonisingdesign.com.[40]

In a similar vein, design historian and scholar Victor Margolin, in an influential article titled "Teaching Design History," advocates a shift away from Eurocentric, modernist approaches to design history and toward a truly global approach that includes design practices from Latin America, Africa, and Asia. He cautions against sprinkling "non-Western" design objects on top of an already existing Eurocentric curriculum, and argues that "design is no less than the conception and planning of the artificial world. Its products include objects, processes, systems, and environments; in short, everything."[41] Margolin feels that an emphasis on rethinking historical narratives to center formerly marginalized or erased design practices, rather than simply including designed objects from more cultures, can help avoid this pitfall.[42]

Teaching to Technologically Transgress In Black feminist author and educator bell hooks's classic text *Teaching to Transgress*, she argues for feminist, antiracist, class-conscious education as the practice of freedom. For hooks, the primary goal of education is for both teacher and students to develop our capacities to think critically, and to take action to transgress boundaries of race, class, and gender. Educators, hooks argues, must recognize ourselves as embodied subjects in the classroom, rather than pretend that we speak from a disembodied place. This acknowledgment of the body brings race, gender, class, and disability explicitly into the pedagogical environment. For hooks, teachers can become aware of, and challenge, our own positions in the classroom and our own tendencies to reproduce relationships of domination. She also calls on educators to discuss and work through racism and sexism in the classroom, rather than plaster over tensions that emerge in student conversations about race, class, and gender, in mistaken efforts to focus on the "real" learning goals of the course material. She emphasizes that curricular revisions are not the only component of liberatory pedagogy: "Once again, we are referring to a discussion of whether or not we subvert the classroom's politics of domination simply by using different material, or by having a different, more radical standpoint. Again and again, you and I are saying that different, more radical subject matter does not create a liberatory pedagogy, that a simple practice like including personal experience may be more constructively challenging than simply changing the curriculum."[43] Thus, design justice pedagogies are not only about revising design curriculum to include more texts by women and femmes, by Black, Indigenous, and other people of color, by LGBTQ and Two-Spirit folks, and/or by Disabled people, although such revisions are certainly necessary. Nor is it sufficient to simply include critical texts about how design often reproduces racism, sexism, or other aspects of the matrix of domination. Instead, design justice pedagogies must support students to actively develop their own critical analysis of design, power, and liberation, in ways that connect with their own lived experience. Educators also must find methods to help students challenge their own ideas about themselves, their relationship to design partners, and the role of design in the world.

What does all this look like in practice? In the next section, I draw from my own teaching experience to explore key challenges to design justice pedagogies.

Lessons from the Codesign Studio

This chapter opened with a vignette of students reflecting on systemic inequality after attending a community organizing meeting at City Life/Vida Urbana in the context of the MIT Codesign Studio.[44] The topical focus of the Codesign Studio, and the community partner organizations, changes each time I teach it.

For example, in 2014, the course focused on surveillance and privacy. Teams worked to design countersurveillance projects, grounded in the needs of communities most heavily targeted by state, military, and corporate surveillance. Projects and partner organizations included *SpideyApp*, an Android-based Stingray detector, with the American Civil Liberties Union of Massachusetts and the Guardian Project; graphics for the *Surveillance Self-Defense Guide*, with the Electronic Frontier Foundation; *I Am Not A Dot*, a project about the sex offender registry, with Citizens United for Rehabilitation of Errants; IPVTech, a research portal about the use of mobile technology by perpetrators of intimate partner violence, with Transition House and The Tor Project; the *UYC SMS Survey Initiative*, an SMS survey system to gather data about students and their experiences of surveillance and police abuse inside New York City high schools, with Urban Youth Collaborative; *Infiltrated*, an interactive, web based documentary about federal infiltration of social movements in the United States, with SoMove (the Social Movements Oral History Tour); and *Bedtime Stories*, an interactive documentary microsite that raises awareness about the injustices of the US immigration detention and deportation system by focusing on the detention bed quota, with Detention Watch Network.

In 2016, inspired by the growing conversation about platform cooperativism,[45] the course focused on partnering with CBOs in the cooperative economy. We wanted to help create a pipeline for triple-bottom-line start-ups, built on free and open-source software, cooperatively owned by their workers, to disrupt exploitative models of work in current low-wage sectors. We partnered with four worker-owned cooperatives in the Boston area. With CERO, a cooperatively owned commercial composting company based in Dorchester, we conducted experiments in sales and marketing and produced a social media campaign about the environmental impacts of food waste and the benefits of composting. With Vida Verde, a cooperative of Brazilian housecleaners, we developed an online price quote calculator, an internal calendar system for scheduling cleanings, and an upgrade to the cooperative's website to make it more easily navigable and search engine optimized. With Loconomics, a freelance jobs platform that is like a cooperatively owned version of TaskRabbit, we collaborated on user testing and prototyped improved interfaces for various tasks. With Restoring Roots, a landscaping coop based in Jamaica Plain, we codesigned a transmedia marketing campaign to promote the cooperative's services, as well as the ideas of urban gardening, permaculture, and worker-owned cooperatives.[46]

In 2017, our partners were youth media organizations across the Boston area, and the projects were related to young people's experience of Boston's housing crisis, gentrification, and displacement. We partnered with ZUMIX and the Urbano Project, two youth arts and media organizations in the Boston area, and NuVu Studio, an innovation school for middle and high school students in Cambridge. Codesign Studio students, ages eleven to twenty-six, gathered weekly at the MIT Center for Civic Media to work together while discussing topics central to design justice, gentrification, and transformative media organizing. Projects included *Open Book/Libro Abierto*, a printed and online book containing handwritten and printed texts along with photos of community members; a series of audio interviews about displacement and community in Egleston Square; *East Boston Voices*, a podcast about gentrification and displacement in East Boston; *Homesticker*, a geolocative media project about home and displacement; and *Rainbow*, an interactive art installation in Central Square's graffiti alley about Cambridge residents' experiences of gentrification.

All the Codesign Studio project teams produce case studies; these can be found at https://codesign.mit.edu/projects. In the case studies, design teams are responsible for reflecting on and critically evaluating their own work. They describe the project context, analyze their design process and the designed object that they produce, and end by discussing key challenges. Over the past six years, in their self-evaluations the codesign teams repeatedly identified the following common challenges: structural inequality can be identified, but not solved through a design process; it's very difficult to define *community* and to operationalize community accountability; it's important to consider various kinds of impact, including how to "do no harm"; it's important to prototype early, and get those prototypes into the hands of community members; broader power dynamics continue to exist within design justice teams; there are significant coordination and logistical challenges to effective community participation; community-facing events are key sites for an inclusive process; it can be hard to ensure clarity about project ownership; and there are sociotechnical constraints on project implementation outside of harmful existing systems. Below, I have organized a discussion of each challenge, with examples drawn from student case studies. I have placed them in dialogue with the Design Justice Network Principles that were introduced at the beginning of this book.

Principle 1: We Use Design to Sustain, Heal, and Empower Our Communities, as Well as to Seek Liberation from Exploitative and Oppressive Systems

The first principle of the Design Justice Network encourages designers to not only critique oppressive systems but also participate in active healing and community empowerment. In practice, student design teams wrestle with the fundamental tension that structural problems identified during design justice research cannot be easily designed away.

Especially in an educational setting, this tension can easily leave student designers feeling overwhelmed, hopeless, or paralyzed by the seeming futility of design work. Although true for all design approaches, it is especially crucial in design justice to find specific ways for participants to feel a sense of completion. Otherwise, the approach may dissuade, rather than encourage, people in marginalized positions within the matrix of domination from participating in design. For example, chapter 3 focused on the need to change exploitative narratives as part of the design process. The CL/VU Change the Game project team, described at the beginning of this chapter, noted tension between deep engagement with questions of structural inequality and the production of concrete product deliverables. They found that a "major challenge was balancing a nebulous concept like 'changing the housing narrative' with needing to produce a concrete product deliverable."[47]

Master narratives, by definition, are very powerful and are difficult to disrupt. The SpideyApp team felt that their biggest challenge was "overcoming people's preconceived ideas about privacy and educating them about the problem and why they should care." This team was frustrated by state narratives about the necessity of surveillance, as well as by many people's sentiments that they "have nothing to hide."[48]

Design justice is a method that centers structural and institutional analysis of power inequality and is interested in root causes, unlike many design approaches. However, even while recognizing that design often can only contribute in limited ways to challenging oppression, it is also a method that's meant to produce real designed objects, interfaces, services, and so on. There is thus an important tension within a design justice approach between dealing with the larger, long-term forces of structural inequality and the need to make something concrete in the here and now that can contribute to sustaining, healing, or empowering a community.

Principle 2: We Center the Voices of Those Who Are Directly Impacted by the Outcomes of the Design Process

Although it is important to be guided by the principle "nothing about us without us," in practice design teams often wrestle with real-world implementation of the second design justice principle. In chapter 2, we discussed the crucial question of "Who gets to do design work?" Especially in design teams that include students, many ask some version of questions like: What is the community in this project? Who gets to speak for the community? How do we make our design process truly accountable?

In the Codesign Studio, the teaching team provides scaffolding to address these questions, primarily by seeking out CBOs that have established track records of doing good work in their communities. We also secure resources to enable CBOs to fully participate in the design process. Organizations typically choose one or two staff members and/or highly engaged community members to participate in the Codesign Studio as project leads; these individuals attend weekly course meetings as well as design workshops and project team meetings. In this way, we work to break down the traditional expert/client relationship, as well as the walls of the classroom. Project teams include students, MIT staff, and staff and volunteers from community partner organizations who all work to design projects together. This approach avoids the dynamic of student designers entering a community that is not their own in search of individual community members to participate in a design process that they (the students) have initiated and conceived.

If at all possible, educators who teach or facilitate design justice courses should find ways to resource community partners. It takes a lot of time and energy to remain engaged in a design process, time that nonprofit staff or social movement organizations may not have. Although community partners may express desire to participate fully in the design process, they are often strapped for resources and understaffed, and staff may have multiple roles and responsibilities. If the design process unfolds over any significant length of time, early enthusiasm may give way to the realities of ongoing work, shifting priorities, and the need to respond to larger developments, crises, and/or political opportunities in the broader landscape.[49] Finding ways to compensate community partners for their time on the project can help mitigate these challenges.

Although bringing CBOs into the design process from the beginning is a key accountability strategy, in the Codesign Studio we also know that a team that is trying to practice design justice needs to develop very clear, transparent, and explicit decision-making processes. One way to do this is to require a written working agreement or memorandum of understanding among all team members. This kind of document describes who is participating, what their respective roles will be, how decision-making will work, ownership of any outputs, and so on.[50] The point is to make the process explicit and clear to all participants. A written agreement is a key starting point, but teams also often need to check in about how their decision-making process is working, as well as about how they feel about the design product(s). This approach is also recommended by the Boston Civic Media Consortium, which examined community-academic partnerships across Greater Boston in 2016 and found that such agreements are often crucial to help mitigate the asymmetrical power relationships between universities and CBOs.[51]

For example, the ZUMIX Codesign Studio team members reflect on this dynamic extensively in their case study. They note that in the process of developing a written MOU, they were "forced to think about not only what was feasible for the end-product, but also to think critically and openly about the planning, decision-making, and implementation processes that this project entails. Who gets to decide which project we choose? Who participates in designing and building the final product?" However, they go on to say that although they had clarity on paper, in practice, representation and accountability became more complicated: "According to our MOU, decision-making powers lie with the

'the project partnership team, composed of ZUMIX staff, the ZUMIX youth representative, and CMS.362 students.' However, our youth representative was never officially selected, leaving this spot empty and, should conflict have arisen, could have left the youth DJ voice silent in the decision-making process. This lack of follow-through was likely the result of inexperience, a desire to 'get things done,' and unrecognized ageism on the part of the core design team."[52]

The team members go on to describe a conflict about what form factor to use in the physical housing for an internet radio device that they built together. The students wanted to laser-cut an acrylic casing in the shape of a giant Z, to represent ZUMIX; the organization staff wanted to house the internet radio in a repurposed wooden old-time radio, to represent the values of remix and sustainability. The students pushed back and created the laser-cut casing, and ultimately, the organization was not satisfied by the project outcome. In their evaluation of challenges, team members reflected that it might have helped to check in regularly about how the decision-making process was working out, rather than just sign a written MOU at the beginning, and then move on to focus primarily on product discussions during team meetings.[53]

Principle 3: We Prioritize Design's Impact on the Community Over the Intentions of the Designer

In design justice pedagogies, educators need to consider not only the learning outcomes for students, but also individual, organizational, and community-level impacts on partners. It's hard to overstate the importance of honestly asking: "What will community members get out of the process?" In particular, community members who live at the intersection of multiple forms of oppression often don't have free time to dedicate to a design process. Ideally, they will be paid for their time, but even so, community partners can sometimes be, and feel, used by the design process. In the worst case, community partner organizations are used by student design teams primarily as a way to access vulnerable populations in order to test student project ideas.[54]

Mistakes and failure are part of learning. However, the start-up discourse that valorizes failure can be particularly harmful in design justice processes. Start-up ideology, such as "move fast, break things" and "fail hard, fail fast," can become a justification for working styles that replicate broader structural inequality, when privileged student designers get to have a learning experience that involves making mistakes in the real world at the expense of community partners.

In the Codesign Studio we have learned that it is important for design teams to think concretely about the kinds of impact they want their projects to have beyond raising awareness.[55] Issue visibility is not enough; project teams also have a responsibility to point people to specific actions they can take, and especially to connect them with existing organizations. For design projects to have large-scale impact, if that is one of the goals of the team (and it need not always be), institutional partnerships are often necessary. In addition to CBOs and networks, government, educational, arts, and media institutions are all possible partners that can bring additional resources to the table, heighten visibility, and scale impact. However, anything that requires institutional approval takes place in a time frame that doesn't usually fit well within an academic calendar. For example, team NuVu wanted to install a public interactive sculpture, but needed to do so without passing through a lengthy process of city approval.[56] Institutional partnerships also introduce additional challenges for design justice work, such as project attribution, control, and ownership.

One key mandate for design justice practitioners is to *do no harm*. Operationalizing this principle in a learning environment can be complex and challenging. In some cases, educators and/or community partners may need to veto student ideas because they would potentially place people from a vulnerable community at risk of harm. For example, in the Urban Youth Collaborative project, design candidates initially included public social media campaigns to document police abuse against high school students in New York City schools. However, UYC organizers reminded MIT students that this approach would place high school students at risk of retaliation from in-school police officers, with whom they have to engage daily.[57]

Principle 4: We View Change as Emergent from an Accountable, Accessible, and Collaborative Process, Rather than as a Point at the End of a Process

Codesign Studio participants often reflect that decision-making in any design project involves a delicate balance between the desire to be inclusive, collaborative, and accountable, and the need to get things done. In many cases, perhaps counterintuitively, most participants feel better about the process if decision-making is constrained to a limited and specific number of moments. For example, especially in early-stage design projects, where the goal is to go from ideation to a prototype, everyone may feel better if feedback is limited to particular rounds rather than constant, ongoing back-and-forth about small requested changes. Limiting the number of rounds of feedback (say, to three) can function to help focus and prioritize the most essential changes between one iteration and the next. This is also the case from my experience across many different kinds of design processes, including projects with a traditional client/designer relationship. For example, Design Action Collective, a worker-owned cooperative that does graphic design and website development, includes a detailed process roadmap that specifies the number of feedback rounds in its boilerplate contracts.

Getting a prototype in front of real-world users early on in the design process is fundamental to making design more accessible. This is crucial because it helps to validate assumptions, reveal faulty thinking, and allow the team to iterate on the selected concept. This is widely understood across many approaches to design. When student teams spend too much time researching, theorizing, analyzing, and ideating, but fail to move quickly enough to mock-ups or prototypes (depending on the type of project), they lose invaluable opportunities to iterate on the project based on user testing and feedback.[58]

It's very easy for design teams operating on a semester schedule to run out of time. To take another example, *Open Book*, developed with Urbano Project to "share the stories of the activists and residents who are intervening in the gentrification of Boston neighborhoods and the displacement of its denizens," produced a compelling prototype object but ran out of time to implement the community partner's ideas about how to use the object to spark public dialogue. They had hoped to bring the book to Boston City Hall for a public event and gather more organizations from across the city to collaborate on content production, but never did so due to time constraints.[59] The Urban Youth Collaborative team had a similar experience: "Too much time was spent investigating options, rather than settling on a platform and tailoring it as needed."[60] Narrowing down from big concepts to working prototypes within the available time can be very difficult. Part of the educator's role is to guide teams through this process with clear expectations and firm deadlines.

Because design justice pedagogies emphasize a balance of process and product, rather than simply valuing "final" products, regular assessments of student work and of the design process can help improve the overall experience for everyone. Leaving assessment to the end of the process, or just to one or two key moments (such as a midterm and the end of the semester), is a mistake. To further complicate matters, students do not always appreciate pedagogy that emphasizes process, real-world contexts, challenges, and partnership; instead, many desire a design studio that allows them to freely explore the limits of their creativity, with evaluation based on a final product.[61]

Principle 5: We See the Role of the Designer as a Facilitator Rather than an Expert

Broader power dynamics do not magically disappear within design teams just because everyone involved is committed to design justice principles. Gender, race, class, disability, education, language, and other forms of structural inequality are always active in educational environments. These forces are in play between students from different backgrounds, between students and educators, between students and community members, and so on. These are complex dynamics that can be difficult to navigate.

Privilege and power never go away, but a design justice studio can become a place where they are explicitly recognized, acknowledged, and discussed. In developing a critical pedagogy of design justice, the facilitator must work to ensure that participants discuss privilege and power, introduce team working agreements that make these dynamics explicit and specify how they will be dealt with, and otherwise make the design process a place for mutual learning and growth around how to challenge the reproduction of structural oppression. There are specific training resources, such as the AORTA anti-oppression training manuals, that can be very helpful with this aspect of design justice work.[62]

It's very difficult to break down the walls between students and community members, although this is one of the goals of design justice pedagogies. At the same time, while we want to destabilize othering and encourage shared connection across various kinds of difference, we don't want to "erase" differences or pretend that they do not exist. To hold these two goals in balance—to break down barriers and create a space of mutual empathy and solidarity, while recognizing and respecting the validity of different standpoints and life experiences—is one of the core challenges of any pedagogy of design justice. Additionally, many students subscribe to a liberal democratic theory of *multistakeholderism*, a concept that has been carefully critiqued by feminist ICT scholar and activist Paula Chakravartty.[63] For example, rather than work closely with a community partner organization that was already actively fighting displacement, one group of students expressed a desire to include people involved in promoting gentrification, such as developers, landlords, and gentrifiers, in their design process.

Besides the different standpoints of people on the design team, most people, including students and community partners, also are used to operating within a client/designer relationship. Students in the process of professionalization who have certain kinds of skills, especially software development, graphic design, or industrial design skills, are often unreceptive to the idea that in a codesign process, they might not be the only "expert" at the table. Although they may have specialized knowledge that the community partner does not, some students are unable to fully respect that the community partner also has specialized knowledge. Some kinds of knowledge are valued much more than others, and students often have internalized a value system that places their own skillset and experiences above those of community organizers and community members.

These dynamics are even more prominent in professional design contexts. In other words, many designers have a highly specialized skillset and value their own skills and opinions more than those of a community partner (or a client). That said, it is worth questioning whether in any particular design justice process it makes sense to challenge the fundamental idea of the client/designer relationship to attempt to create a shared and mutually accountable codesign team, or whether it in fact makes sense to have very clearly articulated client and designer roles.[64] Either way, it is crucial to spell out all roles, responsibilities, and decision-making processes. As feminist scholar Jo Freeman notes in her classic article "The Tyranny of Structurelessness," too often the pretense of a flat structure serves primarily not to truly flatten power dynamics, but simply to mask them.[65]

Even when teams are explicit about their decision-making process, most people are not used to democratic decision making. Throughout our lives, and especially when we are still young people, we are socialized into authoritarian decision-making structures. Most classrooms, workplaces, and families are structured with hierarchical power. Because design justice focuses on fair and meaningful participation in design decisions, one of the goals of design justice pedagogies is to explore the possibility of more democratic decision making within design processes. Student teams usually need significant scaffolding and support for how to do this. For example, team CERO said, "The decision making process was messy because we wanted everyone to sign off on a project before we dove in. … Ultimately we believe that muddling through ideas in this way was useful despite being time-consuming, because we were able to hone our priorities together as a team."[66]

To summarize: privilege and power do not magically disappear in a design justice process. Student designers often expect to operate within a client/designer relationship; also, like most people, they are not used to democratic decision making. Students often subscribe to mainstream ideas about design, and constantly make assumptions about communities that may or may not be true. Some strategies to mitigate these challenges include creating clear, written agreements about project ownership and decision-making processes, as well as validating assumptions early and often.

Principle 6: We Believe that Everyone Is an Expert Based on Their Own Lived Experience and that We All Have Unique and Brilliant Contributions to Bring to a Design Process

The principle that everyone is an expert based on their own lived experience is a crucial element of design justice, but in a design class composed of students, community partners, and support staff, the logistics of inclusion are often quite challenging.

Diverse design teams present special difficulties. It's hard enough to match students with each other, let alone with community partners; it's important to find a good way to match student skills and interests with community projects. One approach, perhaps the most effective for doing real design work, is for students in such a class to already have relationships with the community partner organizations. On the other hand, that approach is limited to those students who already have such relationships, but fostering these relationships is itself an important goal of critical design pedagogy, as noted by scholars, artists, and data scientists Catherine D'Ignazio and Lauren Klein in their approach to teaching feminist data visualization.[67]

Also, teams change: participants often shift during the course of the project for a variety of reasons. This is true for any design process that extends for any length of time but is especially likely when partnering with smaller CBOs. For example, team CERO had a new cooperative worker join at midterm, and that shifted the focus of their project "to collecting more information ... instead of developing an MVP related to the information we already had." For the Loconomics team, one member dropped out, while another was hospitalized with a broken leg.[68]

Even when teams are solid, coordination is tricky. For example, the *Claro Que Si* team found that it was very difficult to coordinate when some team members were students, some were nonprofit staff, and some were working-class people.[69] Scheduling time to work out of class can be "a nightmare";[70] the CERO team constantly struggled to find meeting times that were accessible to both students and CERO workers/owners.[71] In the Codesign Studio, we find that teams need to be reminded specifically to organize a persistent communication channel, whether an email list, chat group, or something else. Teams also need to choose tools for project work. In some cases, it may be best to let teams choose their own working toolset; in others, the educator may want to standardize the tools across the class. The benefit of using the same toolset is that class participants may provide informal peer-to-peer support with the tools; however, teams working on very different types of projects, or with very different kinds of communities, may find a particular tool an imposition. Digital tool selection also tends to privilege the most tech-savvy team members.

Language may also be a barrier. Team Vida Verde experienced difficulties because the home cleaners they worked with were native Portuguese speakers, while the students were mostly monolingual English speakers.[72]

Geographically distributed teams are especially hard to work with. For example, the Urban Youth Collaborative team found communication and coordination very difficult within a team distributed among Cambridge, Boston, Wellesley, and New York City.[73] The Detention Watch Network team noted that when possible, it works best to organize times for distributed teams to meet face to face (ideally) or remotely (if necessary) to sprint on a project together.[74]

Regardless of the amount of scaffolding provided by the educator/facilitator(s), as the CERO project team says, "design processes can be messy and confusing."[75] Indeed, education scholar Brent Mawson has argued that the linear design process models so frequently taught in design classes generally fail to reflect the nonlinear strategies that are actually employed by learners.[76] Learning how to successfully navigate the "messiness" of an inclusive design process that takes everyone's lived experience seriously is ultimately one of the key goals of design justice pedagogies.

Principle 7: We Share Design Knowledge and Tools with Our Communities

Part of the facilitator/educator's role is to support students to engage with community members beyond the classroom walls. As discussed in chapter 4, the physical sites where we choose to engage in design processes have important implications for who is able to participate. For example, it may be possible to participate in existing community events, to move ongoing design meetings to community spaces, and/or to organize community design workshops related to the project.

Often, design teams can piggyback on existing community events to test ideas, gather feedback, and produce content. This was the case for the *Bedtime Stories* codesign project with Detention Watch Network. The design team used a #Not1More immigrant rights event in Jamaica Plain as a site to seek volunteers and shoot key video content for the microsite: "We showed up to this event with our camera gear and a cheap bed, unsolicited, and asked the organizers if they'd be interested in posing for our GIFs. We were able to produce 7 GIFs that day."[77] Another team, Peas in a Podcast, said that they wished they had moved their entire podcast production process into a community radio station.[78] Moving aspects of the design and production process out of the professional design studio, university, or lab and into accessible community sites is a key component of a design justice approach.

Design teams may also organize fun, engaging events, like the DiscoTechs discussed in chapter 4, to bring more community members into the process. However, too often such events are framed as places to make or hack new things, not as places where community members can help generate ideas, make decisions, guide the design process, test out prototypes, or provide meaningful feedback. For example, the Loconomics team members struggled all semester to find appropriate users for testing prototypes, until finally they were able to test with many people at once in the co-op DiscoTech event. Overall, events organized by community partners can be excellent opportunities for many aspects of design, including ideation, testing, validating assumptions, decision making, and more.[79]

Principle 8: We Work toward Sustainable, Community-Led, and Controlled Outcomes

In my experience, many students have been socialized into entrepreneurial neoliberal subjectivity, as articulated so lucidly by scholar, designer, and digital worker advocate Lilly Irani.[80] They often arrive to the classroom primed to believe in and desire individual intellectual property, product ownership, and patents on their work. Universities also increasingly provide support to their students (and faculty) to take the outputs of shared design processes and use them to launch start-up for-profit companies. Students may not have been exposed to conversations about why the commons is important, why free software is important, or why it may make sense for a community-based organization to have ownership of design outcomes. Therefore, in addition to ongoing conversations about these ideas within the space of the classroom, within design justice pedagogies it's essential to create concrete agreements about project ownership and handoff.

Clear, signed agreements about ownership are key. Design projects produce a wide range of outputs: from physical and digital artefacts and objects to working code, from applications installed on particular servers to images and representations of what the project was about, from slide decks, zines, and academic papers to data produced by community partners and community members. All of these are ideally covered in written MOUs. For example, if a project generates data, that data must be shared back with the community partner. However, even when such written agreements exist, it is unfortunately entirely possible that end-of-project transfer of relevant materials never takes place. This typically happens not out of intentional noncompliance, but because after the end of the school year, students move on. For example, in the Codesign Studio, community partner DS4SI had this experience with a student design project meant to capture neighborhood resident views about the possible future of urban planning: "Directly following the project ending, we were all in good spirits. The [Upham's Corner Input Collector (UCIC)] had been made and made beautifully, matching the design and feel of the exhibit and had been there to collect necessary data. Even though it broke down and only worked for about half of the week-long exhibit, we felt positive feelings. A month later however, with none of the data synthesized or even sent back to us in a raw format for us to work through, this project seems much less beneficial."[81] This project partner went on to suggest that it may be crucial to include project handoff to the community partner within the time frame of the semester and to link it to student grades. Also crucial here is the lesson that, when student design teams deploy projects in the real world, they often forget about the need to plan for project maintenance after the semester's end.

Principle 9: We Work toward Nonexploitative Solutions that Reconnect Us to the Earth and to Each Other

This principle may be one of the hardest to realize in practice, in part because of the extensive sociotechnical constraints on its implementation. Design teams always face technical challenges on the path toward realization of their ideal vision; indeed, as discussed in chapter 1, this is a key part of the nature of design. Design justice practitioners who hope to avoid solutions that damage the Earth or that rely on exploitative labor relations face additional layers of constraint on the range of possible options. Of course, no solution is ever perfect, regardless of the criteria, and design can be seen as a permanent *striving toward,* an ongoing process of ideation, iteration, and revision toward the ideal. In design justice pedagogies, understanding this can help mitigate disappointment.

Even when the design team hopes to develop nonexploitative solutions, organizations and individuals are often locked in to particular infrastructures, tools, platforms, or ways of working. For example, Facebook is seen by many design justice practitioners as highly exploitative of user data and as potentially harmful to social movements, but it is also used by most community-based organizations as a key element of communicative practices, so it cannot be ignored by design teams working on a communication campaign.

In some cases, a design project may provide impetus for organizations to shift away from suboptimal or harmful tools and platforms. However, more frequently the design team will have to respect this constraint and adapt the project accordingly. For example, for the EFF Surveillance Self-Defense project, the organization chose to use a content-management system that has a notoriously steep learning curve and that the design team had no familiarity with; this left them dependent on theme-integration work from a third-party developer.[82]

Principle 10: Before Seeking New Design Solutions, We Look for What Is Already Working at the Community Level, and We Honor and Uplift Traditional, Indigenous, and Local Knowledge and Practices

Finally, in design justice pedagogies, it is often the educator/facilitator's role to encourage design teams to first consider what already works at the community level, and to steer students away from the pitfalls of tech solutionism and technochauvanism, as described with such clarity and wit by scholar and data journalist Meredith Broussard.[83] This includes exploring whether the design team might be able to amplify, remix, or otherwise repurpose existing projects, practices, applications, or tools, rather than build something new. Creatively repurposing freely or cheaply available elements is useful for rapid prototyping, idea validation, cost reduction, long-term sustainability, and more. However, there is no magic bullet. Building something new, on the one hand, or repurposing existing tools or products, on the other, both bring their own challenges. The desire to build something new may keep project teams from using existing tools that might be "good enough" to implement the project, if not a perfect fit. At the same time, limitations of existing products may make it very difficult to implement the project vision, and existing tools, platforms, and infrastructure often violate Principle 9.

One good way to navigate these challenges is for the educator/facilitator to guide the project team to implement a mockup or rough prototype of their top design candidate, using already existing tools, early on in the process. For example, the Urbano team case study describes the twists and turns of trying to implement a stop-motion animation studio and screening room in a small suitcase, using a prepaid mobile hotspot, Vine, and dual Kindles: "We spent significant time trying to find a way through the walled garden of Vine … we quickly found that the only tablets that run Vine are Kindles."[84]

Finally, one of the most crucial, if seemingly obvious, lessons here is the value of students spending time with the community. In a design justice process, it is crucial for the whole team to physically spend time with the community that is supposed to lead the process. As DS4SI noted, "We realized the importance of the students coming to Uphams Corner too late."[85] The Neighbormedia team also felt it was very important to meet in the community partners' space,[86] as much and as early as possible in the design process.

Conclusions: Learning to Code as Liberatory World-Making, or Workplace Preparedness under Neoliberal Technoculture?

In 2016, the Obama administration announced a Computer Science for All initiative and proposed $4 billion for states, $100 million for school districts, and $135 million for the National Science Foundation and the Corporation for National and Community Service to train computer science teachers. President Obama announced the program with the following statement:

> We live in a time of extraordinary change—change that's affecting the way we live and the way we work. New technology replaces any job where work can be automated. Workers need more skills to get ahead. These changes aren't new, and they're only going to accelerate. So the question we have to ask ourselves is, "How can we make sure everyone has a fair shot at success in this new economy?" … I've got a plan to help make sure all our kids get an opportunity to learn computer science, especially girls and minorities. It's called Computer Science For All. And it means just what it says—giving every student in America an early start at learning the skills they'll need to get ahead in the new economy.[87]

Most of the money never materialized,[88] but the underlying assumptions have only gained power. Teaching people how to code is increasingly presented as a key goal—perhaps *the* key goal—for the education system under late-stage informational capitalism. Producing a workforce with software development skills or otherwise ensuring a sufficient supply of workers with the ability to code has become a key national project for many countries.

However, the US educational system exists under conditions of prolonged imposed resource scarcity. Funding for public education is under constant attack, and children are funneled into a two-tier educational system. The lower tier is a warehousing and feeder system for the prison industrial complex, known as the *school-to-prison pipeline*.[89] In the broader context of rising wealth inequality, a winner-take-all dynamic is at play, with wealthy white people withdrawing their children and tax dollars from schools that used to serve mixed-income and multiracial populations. Forty-three percent of Black and/or Latinx students attend schools with poverty rates above 80 percent, compared to 4 percent of whites.[90]

Schools in low-income communities of color are rarely allocated the resources they need to provide high-quality STEM education. As a result, Black, Latinx, and/or low-income students are more likely to be taught by less experienced teachers, receive less funding per student, face lower expectations, and score lower on standardized STEM tests, and are less likely to enter STEM fields in higher education.[91] School pushout and in-school abuses faced by LGBTQ and GNC youth, especially LGBTQ youth of color,[92] are additional factors that militate against more women, POC, and LGBTQI people gaining STEM education and thereby moving into coding, design, and technology professions. What's more, under the austerity conditions of radically underfunded public education, in public schools learning to code ends up positioned against other skills, especially humanities and the arts. Budget cuts come first for subjects that emphasize creativity and critical thinking.

Meanwhile, private and wealthier schools increasingly provide a plethora of computation and design courses. At the same time, as sociologist Tressie McMillan Cottom documents in her book *Lower Ed: The Troubling Rise of For-Profit Colleges in the New Economy*, for-profit universities that promise to teach coding skills and secure jobs for their graduates proliferate both on and offline.[93] Many of the most visible for-profit coding schools and boot camps are expensive, inaccessible, and have dubious placement outcomes.[94]

Unsurprisingly, given this context, digital learning among young people remains structured by race, class, and gender. In a recent study of digital learning, education researchers Mimi Ito and Justin Reich find that, in many cases, digital learning technologies such as MOOCs and online courses, in-school computing classes, and other interventions actually exacerbate inequalities in learning outcomes between low-income and wealthier students, between students of color and white students, and between male and female students. In addition, they note that the use of digital technology in education often unintentionally reproduces inequality, in large part due to "institutionalized and unconscious bias and social distance between developers and those they seek to serve."[95]

Ultimately, as more and more production processes are digitized and as design becomes primarily dependent on software, there is a growing design education gap. In other words, although the digitization of design theoretically democratizes design education, in practice it disproportionately benefits already powerful groups. The benefits of design education remain structured by the matrix of domination.

Democratizing Design Education

Even under these extremely difficult conditions, there is no paucity of brilliant, innovative individuals and organizations from marginalized communities who focus on the democratization of design education. Largely due to their efforts, some of the goals of pop ed and other liberatory design pedagogies presented in this chapter are arguably becoming mainstreamed.

For example, learning to code is increasingly taught in ways that emphasize diversity, creativity, and critical thinking. This is especially taking place in K-12 education. Educators Jane Margolis and Joanna Goode, authors of *Stuck in the Shallow End*, received NSF funding to develop the Exploring Computer Science curriculum, a year-long introduction to computer science for high schoolers, as well as a teacher professional development program with an emphasis on increasing equity in computing.[96] This curriculum has been widely adopted. Code.org, a large nonprofit that focuses on expanding computer science education in schools and reaches 30 percent of K–12 students in the United States, explicitly works to increase the participation of women and underrepresented minorities in computer science, with some promising outcomes in high school CS course participation rates.[97] The MIT Teaching Systems Lab has developed a research-based approach to dealing with bias in teaching, and some of their findings have been incorporated into Code.org's approach.[98] Scratch, the widely used platform for computational literacy that was developed at the MIT Media Lab's Lifelong Kindergarten group, is entirely focused on creative computing.[99] There is also a new wave of attention in higher education to teaching computer science in ways linked to social and ethical concerns.[100]

There has also been a recent increase in attention to (and funding to address) the lack of diversity in STEM education, as well as continued efforts by many organizations that have long worked toward gender parity and racial equity in STEM. For example, the National Center for Women and Information Technology (NCWIT), a community of several hundred companies, universities, government agencies, and nonprofit organizations, was founded in 2004 by the National Science Foundation to advance women and girls' participation in ICTs.[101] Alongside longstanding initiatives, newer organizations that focus on building the design, tech, and media skills of girls and women, B/I/PoC, and LGBTQ folks also continue to emerge. Media scholar Christina Dunbar-Hester describes Debian Women, Geek Feminism (geekfeminism.org), PyLadies, Genderchangers (https://www.genderchangers.org), and other groups that, since the early 2000s, have focused on increasing the participation of women in free software development, within Python, Debian, and Linux communities.[102] Black Girls Code, launched in 2011, teaches young African American women the basics of computer science and software development.[103] Girls Who Code, launched in 2012, focuses on eliminating the gender gap in the technology and engineering sectors.[104] Code2040, based in San Francisco, works "to ensure that by the year 2040—when the US will be majority Black and Latinx—we are proportionally represented in America's innovation economy as technologists, investors, thought leaders, and entrepreneurs."[105]

These kinds of organizations (and there are many more) are doing crucial work. Undoubtedly, design justice requires a broad democratization of software development capabilities. In a world structured by software, dismantling the matrix of domination requires that people from more diverse backgrounds learn coding skills. However, design justice principles impel us to also ask: Will all this coding education necessarily advance our collective liberation? How can we ensure that it does?

Make All People Good Coders, or Make All Coders Good People?

A century ago, sociologist, historian, and Black liberation activist W. E. B. Du Bois famously engaged in a sustained debate with educator, author, and presidential advisor Booker T. Washington over the nature of the education system that was to be put in place for Black people after the end of slavery, the collapse of Reconstruction, and the rise of Jim Crow. At the turn of the century, Washington created a system of vocational schools that focused on teaching Black people marketable skills for employability in agriculture and industry. Du Bois, on the other hand, argued for the creation of Black liberal arts colleges to foster a new generation of Black leaders, critical thinkers, cultural luminaries, and (above all else) teachers, who would be able to bring the benefits of education to all Black people.[106] For Du Bois, in a phrase that he would repeat in multiple speeches and writings, "The object of education was not to make [people] carpenters, but to make carpenters [people]."[107] Following Du Bois, we might ask of the recent emphasis on learning to code: Is the ultimate object to make people good coders, or to make coders good people?

The ability to design new technologies, platforms, and systems is undoubtedly a key skill in today's economy, and the democratization of this ability is one key goal of design justice. However, are we satisfied with everyone learning to code, if the end game is to produce (admittedly more "diverse") coders who will primarily work to ensure the continued profitability of capitalist start-ups and technology giants? Or, like Du Bois, might we advocate that people learn to code in ways that also push them to think more critically about software, technology, and design and that prepare them to help reshape technology in the service of human liberation and ecological sustainability, rather than the matrix of domination?

Hopefully, both are possible: design pedagogies that promote critical thinking are not incompatible with the development of practical design skills. Design justice is a framework that can help guide us as we seek to teach computing, software development, and design in ways that support, rather than suppress, the development of critical consciousness and that provide scaffolding for learners' connections to the social movements that are necessary to transform our world.

Directions for Future Work: From #TechWontBuildIt to #DesignJustice

Figure 6.1
No Tech for ICE, from the #TechWontBuildIt campaign.

> To: Talent Acquisition at Amazon. Thank you for reaching out. While I'm sure this would be a great opportunity, I have no interest in working for a company that so eagerly provides the infrastructure that ICE relies on to keep human beings in cages, that sells facial recognition technology to police, and that treats its warehouse workers as less than human. Best wishes, [redacted].
>
> —Anonymous participant in #TechWontBuildIt

Tech workers have recently been building power through active refusal to work on oppressive technology projects, often under the banner of the hashtag #TechWontBuildIt. As Lauren Luo, a student at MIT in my Networked Social Movements seminar, describes it:

> On December 13, 2016, the day before top tech executives met with Donald Trump in a tech summit, a group of tech workers released the Never Again pledge

> … "that they will refuse to build a database identifying people by race, religion, or national origin." Just over a month later, both Google Co-Founder Sergey Brin and Y-Combinator president Sam Altman joined protests on January 28, 2017 at San Francisco International Airport in opposition to President Trump's executive order that banned immigration [from] seven Muslim-majority countries. Two days later, over 2,000 Google employees in offices around the world staged a walkout and donated more than $2 million (matched by Google) to a crisis fund for nonprofit groups working with refugees.[1]

This wave of activity continued to build. By 2018, more than four thousand Googlers organized a campaign for their company to drop Project Maven, a Department of Defense (DOD) contract to develop image-recognition systems for drone warfare.[2] Scholars and scientists expressed solidarity with the workers; Lucy Suchman, Lilly Irani, and Peter Asaro, together with the International Committee for Robot Arms Control, organized an open letter in support of the campaign that attracted over 1,100 signatories, including prominent figures such as Terry Winograd, a computer science professor who was Google cofounder Larry Page's graduate advisor at Stanford.[3] Ultimately, by June 2018, Google leadership announced that it would drop the project.

Throughout the summer of 2018, #TechWontBuildIt grew in parallel with a cycle of immigrant rights protests. #KeepFamiliesTogether mobilizations swept the country after revelations about the Trump administration's policy of separating thousands of migrant children from their parents, along with images of very young children and toddlers locked in makeshift detention centers under awful conditions.[4] Investigative reporters and human rights organizations found that some of these children were drugged against their will, and documented cases of child sexual, physical, and emotional abuse, as well as child deaths in ICE custody.[5] In response, as #KeepFamiliesTogether and #AbolishICE took prominence in the media cycle, Microsoft workers pushed their company to drop a $19.4 million dollar contract with ICE. Together with immigrant rights organizations, tech workers organized #NoTechForICE protests at Microsoft stores in cities including Seattle, Boston, New York City, and more.[6] Following Microsoft's acquisition of GitHub, the largest repository of free/libre and open-source code, nearly three hundred open-source developers pledged to take their projects off the platform unless Microsoft dropped its ICE contract. (As I write these words, Microsoft's leadership has not yet responded, but pressure continues to mount.)

"Decided to respond to a recruiting email for a change today #TechWontBuildIT #NoTechForICE"
–Anna Geiduschek (@ageiduschek)

"I did this the other day too! #TechWontBuildIt"
–(@_ifnotbyfaith)

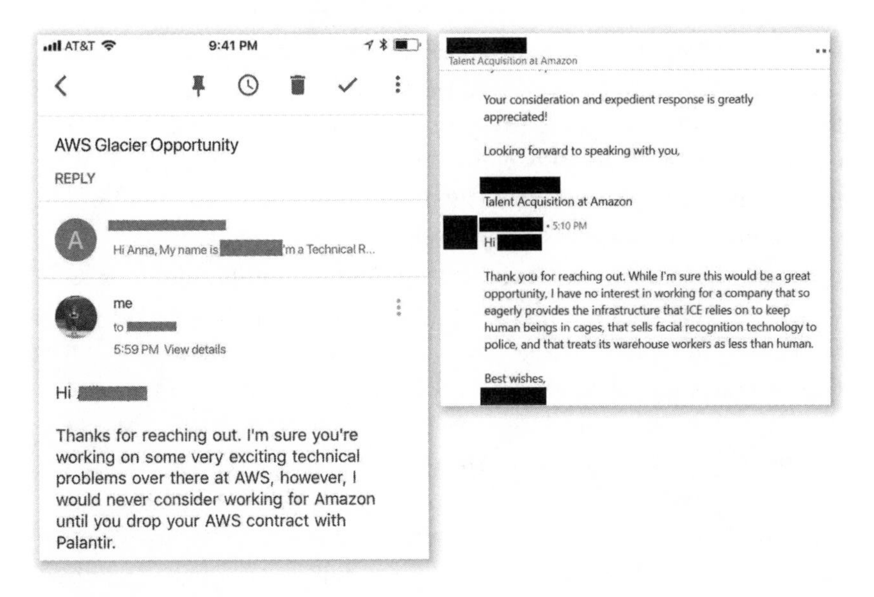

Figure 6.2
Tech workers tweet responses to recruiters from companies targeted by #TechWontBuildIt. *Source:* Screen captures from Twitter.

At Salesforce, workers used the hashtag #CancelTheContract to call for an end to a database services agreement with US Customs and Border Protection (CBP). Hundreds of workers signed a petition, and dozens protested at the Salesforce HQ in the Bay Area on Monday, July 9, 2018.[7] A coffee shop in San Francisco declined a $40,000 vendor contract at Dreamforce, Salesforce's annual conference, in protest of the contract,[8] and RAICES, a grassroots immigrant rights organization in Texas, turned down a $250,000 donation from Salesforce.[9] The management consulting firm McKinsey, one of the most prestigious and powerful in the world, ended a $20 million contract with ICE, although other firms such as Booz Allen Hamilton, Deloitte, and PricewaterhouseCoopers continued to advise ICE on "information systems, data integration, and analytics."[10]

Some tech workers developed a new pressure tactic: they responded to tech company recruiters with notes about their ethical stances, then publicly shared their responses via social media (highlighted in figure 6.2).

Immigrant rights mobilizations were key to the momentum of #TechWontBuildIt in 2018, but tech workers are organizing across many different issue areas. For example, after the leak of Google leadership's internal plans to build a censored search engine for the Chinese market under the name Project Dragonfly, Google workers mobilized again; by August 2018, Google announced that it would not be entering China any time soon.[11] Amazon workers pushed for their company to stop selling Amazon Rekognition facial recognition technology to law enforcement[12] and to stop selling Amazon cloud services to military/intelligence data analysis firm Palantir.[13] Over two thousand IBM employees signed a petition to demand that they be able to opt out of work on government contracts that violate civil liberties.[14] In November 2018, about 20 percent of Google employees worldwide (more than twenty thousand people in offices in more than fifty cities) walked out in protest against the company's sexual harassment policies, among other demands.[15] Although organized tech workers do not always achieve their immediate goals, their recent efforts have contributed to an important expansion of the horizon of possibility.

Through all this organizing, the Tech Workers Coalition (TWC, initially founded in 2016) emerged as a key networked social movement organization. TWC supports the efforts of tech workers across various firms to hold their industry accountable and to transform its practices. Their mission: "Guided by our vision for an inclusive & equitable tech industry, TWC organizes to build worker power through rank & file self-organization and education."[16] Another organization, Science for the People (SftP), expanded chapters to twelve cities and helped organize protests at multiple Microsoft offices in late July 2018.[17] SftP is a present-day reboot of an organization that was initially born in 1969 at MIT, when scientists (faculty, staff, and students) walked out of their labs and joined a university-wide teach-in to protest the militarization of research.[18] The story of the March 4, 1969, events and transcripts of the speeches at the teach-in have recently been republished by the MIT Press in an anniversary edition of the book *March 4: Scientists, Students, and Society*.[19]

#TechWontBuildIt runs counter to the enduring myth that tech workers are apolitical—a myth because there is a long history of politically engaged scientists, technologists, and designers, although the dominant cultural narratives about this kind of work tend to erase that history. For example, the Federation of American Scientists, formed in 1945 by Manhattan Project scientists to limit and control the use of nuclear technology, is still active today.[20] This group influenced a wide range of science and technology policies and is responsible for the creation of the Atomic Energy Commission, among other outcomes.[21] Computer Professionals for Social Responsibility (CPSR), key architects of the long-running Participatory Design conference, were active from 1983 to 2013; CPSR was born hand in hand with the anti-nuclear-proliferation movement of the time and advocated against the use of computers for war.[22] Chapter 1 of this book explored various explicitly political subfields of design and technology in both theory and practice, such as value sensitive design, inclusive design, and decolonizing design. Chapter 3 told the story of how radical techies from the Indymedia network, always tightly tied to the global justice movement, were key to innovations in open publishing, people-powered news, and the massification of DIY media production.[23] Chapter 4 traced the roots of hackerspaces to autonomist and anarchist social centers.

Campaigns to democratize technology are most effective when tied to larger social movements. The book *Grassroots Innovation Movements*, by scholars Adrian Smith, Mariano Fressoli, Dinesh Abrol, Elisa Arond, and Adrian Ely (discussed in chapter 3), provides extended case studies of other examples, including the Movement for Socially Useful Production; the Appropriate Technology Movement; the People's Science Movement in India; makerspaces, hackerspaces, and fablabs; the Social Technologies Network; and the Honey Bee Network.[24] Communication scholar Sandra Braman's work on the history of the Internet provides extensive evidence that the political views of computer scientists and electrical engineers inform their technical design decisions (Braman is the editor of the Information Policy Series at the MIT Press, and this book is part of that series).[25] Most recently, Ruha Benjamin and the JUST DATA Lab have compiled an excellent resource guide to organizations that currently work at the intersection of technology, design, data, and social justice.[26]

Many tech workers, including designers, programmers, and more, are mobilizing today for the same reasons as progressive and radical people in every field in the United States: the openly racist, misogynist, ableist, Islamophobic, anti-immigrant, trans*- and queer-phobic Trump administration demands a response. Since 2016, social movements have grown, built their membership and participant base, organized historic street mobilizations, run candidates for elections (and, in 2018, won key races), taken direct action, and used a wide range of tactics to resist and shift power. Tech workers also feel moved to be part of what social movement scholar Ruud Koopmans calls the current *wave of contention*,[27] or what media and information studies professor Nick Dyer-Witheford calls a *cycle of struggle*.[28]

In other words, tech worker mobilization today is part of both a long history of similar actions and a currently intensified cycle of struggle across diverse, networked movements. The calls to end ICE contracts at Microsoft and Salesforce would not have gained visibility without their ties to immigrant rights mobilizations triggered by the Trump administration's brutal child separation policy. Amazon workers' demands against facial recognition contracts with law enforcement departments would not be so salient without the ongoing #BlackLivesMatter movement, even as the current wave of resistance to surveillance of B/I/PoC has to be understood also in the context of the role of surveillant sociotechnical practices throughout centuries of settler colonialism, white supremacy, and slavery.[29] The unprecedented scale of the March for Science in the aftermath of the Trump election, with its federated structure of local organizing committees, also provided fertile ground for the emergence of #TechWontBuildIt, even as the March for Science was troubled by internal conflicts about the centrality of race, class, gender, migration, disability, and other arenas of struggle.[30]

Designers, developers, and technologists occupy privileged positions in the global economy. Without them, the infrastructure utilized by larger systems of oppression can't be built or maintained. Many of these workers know this and are getting organized to put pressure on their companies and institutions. The #TechWontBuildIt movement members' refusal to participate in the design of explicitly oppressive sociotechnical systems is an important development. Alongside the continued necessity of refusal, the Design Justice Network (and many kindred organizations) seeks to advance the conversation about what it takes to not only resist participation in the design of oppressive systems, but also to design, build, and maintain alternatives. Hopefully, designers in many fields, from big tech to graphic design to architecture and beyond, will find the design justice framework helpful in their efforts to transform design firms, industries, and overall practices.

Questions for Design Justice Practitioners

To effect that kind of transformation, this book has argued that we need to better understand how design reproduces the matrix of domination through varied mechanisms, including the distribution of affordances and disaffordances that we encode into technologies (design values); who gets paid to do design work and who controls design processes (design practices); the stories that we choose to tell about design (design narratives); the inclusion and exclusion of various kinds of people from privileged design locations (design sites); and the methods we use to teach and learn about design (design pedagogies). Throughout, this book has also focused on concrete examples of people, organizations, and networks who are actively doing design justice work today.

Yet there is so much more to do. This final section of the book reflects briefly on a few questions about design justice, and then discusses possible directions for future work.

Further Specification by Design Domain?

First, although most examples in this book are drawn from specific design domains—in particular, software development—I believe the broader questions sketched here are applicable to all activities that fit under the rubric of design. Still, in practice, little can be accomplished without more field specificity. In some design fields, this is already happening. For example, as discussed in chapter 1, there is a growing wave of activity focused on rethinking AI and machine learning through a social justice lens, such as work by research institutes AI Now, Data & Society, and the Data Justice Lab; conferences and networks like FAT*, Data for Black Lives, Black in AI, and the Our Data Bodies Project; and scholars and activists like Safiya Noble, Meredith Broussard, Virginia

Eubanks, Timnit Gebru, Joy Buolamwini, Ruha Benjamin, Mimi Onuoha, Diana Nucera, and many others. We urgently need more critical analysis in every design domain. Hopefully, others will be inspired to extend and deepen the discussion of design justice to various fields, including industrial design, service design, architecture, urban planning,[31] graphic design, fashion, and more.

Tensions between Process and Outcomes?

At the Design Research Society in Limerick in 2018, I presented a paper about design justice that laid out many of the arguments in this book.[32] During the Q&A, one member of the audience said (to paraphrase): "Design justice sounds nice, but it's not practical or possible in real life." To which the only response must be: We have to articulate a vision of the world we want, don't we? Designers who ignore questions about process on the grounds that shipping the product is more important are deploying a version of the Machiavellian argument that the ends justify the means.[33]

Of course, in theory, design justice calls for both equitable design processes and just design outcomes. However, in practice, all design projects have limited resources and limited time, and there are nearly always trade-offs between the inclusivity of a process and the need to ship a product. Put another way, there are real tensions between those design justice principles that emphasize an inclusive design process and those that prioritize impact on a community over the designers' intentions. A design project may be wonderfully inclusive, provide all participants with a sense of ownership, and reward people equitably for their work, but fail to produce a design product that is useful to the community. Alternately, a process that is not at all inclusive may produce a product that is useful and widely loved. These tensions are not fully resolvable, but they also do not invalidate design justice practitioners' attempts to pay attention to both procedural and distributive justice.

Some may believe that the design justice approach makes it impossible to actually design things and release them into the world. If designers spend all our time evaluating the differential impacts of our work on various subgroups of people, they say, we will never be able to complete projects. It's true that design justice practitioners have to take care that critique does not become our primary activity; an overemphasis on testing, evaluation, and critique can indeed be ultimately disempowering. At the same time, explicit critique paired with alternative proposals can be very productive.

What's more, though it may be true that it is technically impossible to consider all the possible ways design reproduces inequalities, the response to "the perfect is the enemy of the good" is "just try!" There may be no perfect design process, where a multicultural, multilingual, queer, variously abled group of designers, researchers, community members, coders, and testers frolic together under a happy rainbow of radically inclusive design (although such a process does sound like a lot of fun!), but there are many, many designers in various fields who work every day to make our processes more inclusive and more just.

A more moderate version of this argument holds that design justice dramatically slows real-world design processes down too much to be viable. In practice, though, a design justice team can work either quickly or slowly, just as in any design approach. In addition, design justice is an approach that will become easier as it matures, as more people practice it, and as specialized domain-specific tools and practices become available. Also, if it is true that a design justice process typically takes more time, perhaps going slower is worth it to build a better, more just and sustainable world.

The Paradox of Pragmatic Design?

Design justice requires that we use a lens broad enough to capture structural inequality, which is not "solvable" in any traditional sense; at the same time, successful design justice projects must produce more than critique. Design produces things: objects, systems, interfaces, apps, illustrations, clothing, machines, buildings, and so on. This is the *paradox of pragmatic design* within a design justice framework: to develop workable designs and to generate products, designers must engage with the realities of limited resources. A radical, utopian design that won't be implemented because it requires resources that aren't available will not improve people's daily lives in the immediate future, whereas a limited, pragmatic design that is organized to meet available resources may be prototyped, revised, rolled out, and in the best case actually provide real benefits to real people. However, if resource constraints become an excuse to avoid examining the root of the problem area, then designers will almost always end up, at best, providing Band-Aids for deep wounds and, at worst, actively serving existing power structures. If we take seriously the idea that current power structures are not only unjust but also steadily leading humanity down an unsustainable path that ends in planetary ecological collapse and species death, then we can't be satisfied with purely pragmatic design.

For example, imagine a design team working hand in hand with a fishing community. The community identifies polluted water as its greatest challenge and as the area for intervention. The pollutants are coming from an upstream petrochemical plant. Should the team allocate its resources to develop and distribute a filtration device that can greatly reduce, but not eliminate, the pollutants from personal drinking water? Or should they allocate the resources toward attempts to make the owners of the petrochemical plant stop polluting the water? Perhaps it's possible to do both at once, with a public campaign targeting the plant and/or regulators while also raising money to build and distribute filters?

Designer and scholar Carl DiSalvo argues for *adversarial design*, an approach rooted in the political theory of agonism. DiSalvo urges designers to create contestational objects, challenge hegemonic power structures, and offer speculative alternatives. As DiSalvo says, "Design can produce a shift toward action that models alternative presents and possible futures in material and experiential form."[34] Another approach might be to engage in parallel pragmatics/utopics: to systematically and explicitly develop radical, or utopian, design solutions within the context of each design project, either at an early stage (during ideation) or alongside and in parallel with the pragmatic design product. The design process itself then becomes an exercise in radical visioning: the design team, led by people from the most directly affected community, explores the root of the problem and develops ideas for systems change, in addition to ideas for products or services that can be implemented within the resource limitations of the project. In this way, design outputs include greater understanding of multilevel problems, proposals for radical transformation, and traditional design outputs. In other words, design becomes part of a praxis of liberation, rather than a tool for, at best, incremental improvement within a context of steadily declining possibilities and, at worst, an extractive instrument for mining ideas from already oppressed communities.

Broader Applicability of Black Feminist Thought?

Another question is: Why should Black feminist thought be used as a foundation for rethinking design as an overall human enterprise? Some may feel this is a move that subjugates other forms of knowledge, or find it to be an undue universalization of a particular understanding of race, class, and gender that centers the US[35] context. However, the core concepts of intersectionality and the matrix of domination, developed by Black feminist scholars and activists, are not themselves unique to the United States. White supremacist heteropatriarchal capitalism and settler colonialism—the matrix of domination—is an ongoing global process, although it operates differently in different places and at different scales.[36] True, the specific implications for design theory and practice must be specified and localized. However, there is no reason why design justice as a framework should only be useful in the US context.

Black feminist thought has increasing influence in design, media, and communication scholarship. In the conclusion to her book *Algorithms of Oppression*, Safiya Noble calls for *black feminist technology studies* (BFTS), "an epistemological approach to researching gendered and racialized identities in digital and analog media studies."[37] The group of scholars with the Center for Critical Race and Digital Studies focus on how "structural inequalities in digital media technologies and systems produce disparate and adverse impacts on communities and individuals of color in the U.S. and across the globe" and also work to "envision the digital as a potential means for generating greater racial empowerment, personal and political agency, democratic participation and activism that diminishes inequalities."[38] Deena Khalil and Meredith Kier have described what they call *critical race design*, an approach to antiracist design grounded in Black feminist thought.[39] André Brock synthesizes critical race theory, Black feminism, and queer theory to propose critical technocultural discourse analysis (CTDA), an approach that centers the "epistemological standpoint of underserved ICT users so as to avoid deficit-based models of underrepresented populations' technology use."[40] Future design justice work might explore each of these kindred approaches' implications for design practice.

In the conclusion to her canonical work *Black Feminist Thought,* Patricia Hill Collins discusses the need to engage power as it operates within what she calls the *structural, disciplinary, hegemonic,* and *interpersonal domains.* Future design justice practitioners will need to engage in each of these domains of power via community-led and accountable design processes. For example, in the structural domain of power, design justice might look like redesigning the institutions of employment, education, housing, health, communication, law, business, and government so that they more equitably distribute benefits and burdens. In the disciplinary domain of power, it might mean resisting and redesigning systems of bureaucracy and surveillance, as well as dismantling the prison industrial complex. In the hegemonic domain of power—the realm of ideology, culture, and consciousness—design justice suggests the need to create new narratives about who has participated in the design of our world so far and who gets to be involved in the future. In the interpersonal domain of power, characterized by everyday acts, small and large, of oppression and resistance, a design justice approach invites us to reconfigure everything from human computer interfaces to the built environment in ways that will more equitably distribute affordances and disaffordances.

Finally, I offer a brief note about why, as a white trans* femme, I personally center Black feminist thought in my attempts to interrogate and extend design theory, when there is a history of white scholars appropriating and erasing Black women's work. I work with Black feminist theory because Black feminists created many of the concepts that (to me) most clearly articulate the dynamics of oppression and resistance. Black feminist thought and the Black women who create it are essential to any liberatory theory and practice. By citing Black women's work throughout this book, I hope that rather than appropriation and erasure, I have contributed in a small way to centering Black feminist scholarship and activism in design theory and practice. I also acknowledge the multiple forms of privilege that I benefit from based on my particular location within the matrix of domination, including whiteness, employment at a powerful university, US citizenship under ongoing settler colonialism, and my lack of lived experience with disability (mostly), even as I face particular forms of oppression as a nonbinary trans* person. I urge readers to further explore the powerful analysis coming from spaces like the Center for Critical Race and Digital Studies, the resources assembled by Melissa Brown at blackfeminisms.com, and the JUST DATA Lab resource guide, gathered by Ruha Benjamin at www.thejustdatalab.com/resources.

The next section describes possible directions for future work, organized according to the top level categories from this book's five chapters: values, practices, narratives, sites, and pedagogies.

Values

Chapter 1 addressed the question, "How do the affordances of the digital objects and systems that we design encode, reproduce, and/or challenge power?" However, technologies frequently, if not always, have unintended consequences.[41] Designers never really know how the things we make will be used. This is a significant area of research in science and technology studies, and it has important implications for any attempts to embed values in sociotechnical systems. To address this question, and for design justice methods to be broadly adopted, we need to develop evaluation criteria, as well as guidelines, standards, codes, and laws, while remaining attuned to the dangers of extractive forms of knowledge production.

Evaluation and Impact Assessment

How might we evaluate design projects according to the design justice principles? One approach is outlined in the *Design Justice Zine*, no. 2. For any design project, we can ask three questions: Who participated in the design process? Who benefited from the design? And who was harmed by the design?[42] The zine provides examples of applying this approach to several recent design projects in Detroit.

In certain types of design practice, formal accessibility evaluations are required by law. For example, the ADA requires compliance with accessibility standards in architectural design, web design, and other domains.[43] As a result, accessibility assessment processes, tools, and metrics, such as web services and browser plug-ins, are widely used by designers to check for ADA compliance. More in-depth compliance testing is also available as a service, with an ecology of firms that are available to conduct audits at various levels of detail. Unfortunately, as design historian and feminist scholar Aimi Hamraie points out, for the most part designers approach accessibility as a post hoc checklist and as a burden required by their company's legal team. Regulation, legal accountability, and mandated accessibility compliance should certainly be seen as real victories based on organizing and disability activism, but they are no panacea.[44]

Others propose formal design discrimination impact assessments based on environmental impact assessments—an approach developed by the environmental justice movement. In the early 1980s, scholar of rural feminisms Corlann Bush suggested gender impact assessment reports for design projects.[45] A small but growing set of firms and organizations, including the Algorithmic Justice League, provide algorithmic accountability audits; Deborah Raji and Joy Buolamwini recently studied the impact of these audits and found that public, multifirm, intersectional algorithmic bias audits do produce improved outcomes in products sold by targeted firms.[46]

Tools to support design justice evaluation include intersectional benchmarks, such as the Pilot Parliament Benchmark dataset created by Buolamwini (2017) to test facial analysis algorithms' ability to classify gender across diverse skin tones; libraries for use by software developers, such as those proposed by proponents of value-sensitive design; how-to guidelines, manuals, and handbooks like those produced by the Design Justice Network and the Detroit Community Technology Project; model working agreements and MOUs, such as those gathered and shared by the Boston Civic Media Consortium;[47] and many others. We also must develop design justice auditing methods that account not only for the intersectional nature of identities, but also for the fluidity of identity categories (which shift over time at a societal level), individual identification (which may shift over an individual's lifetime), and expression/performance (which constantly shifts, consciously or not, in the course of daily life).[48]

Design justice practitioners can expand these types of tools and services to make it easier for more design teams to evaluate for discriminatory design through an intersectional lens. Designers in multiple fields need tools to conduct intersectional audits, and we need to foster an ecology of firms that will audit using design justice criteria. The point is ultimately not to impose a single rubric but to encourage designers and communities to develop and share many different evaluative approaches that are rooted in design justice principles.

Guidelines, Standards, Codes, and Laws

Design justice principles can also be used to produce guidelines, standards, and codes, and designers then need to organize for their adoption by standards bodies and professional associations. Different design domains require different kinds of design justice guidelines. For example, the principles of universal design, compiled in 1997 by universal design advocates Bettye Rose Connell, Mike Jones, Ron Mace, Jim Mueller, Abir Mullick, Elaine Ostroff, Jon Sanford, Ed Steinfeld, Molly Story, and Gregg Vanderheiden, include both overarching principles and specific guidelines for designers to help them implement those principles.[49]

Standards adoption is sometimes voluntary; in other cases, standards become legal requirements. The International Code Committee develops codes for safe buildings;[50] the National Institute of Standards in Technology (NIST) produces standards across a wide range of technological domains in the United States. For example, NIST is currently exploring standards to curb algorithmic bias. In HCI, practitioners of user-centered design are guided by the International Organization for Standardization (ISO) document ISO 9241-210:2010, "Ergonomics of Human-System Interaction—Part 210: Human-Centred Design for Interactive Systems."[51] To take yet another example, there is a tradition of human rights and social justice advocacy at the Internet Engineering Task Force (IETF), the key standards-setting body for the global internet. Lawyers, hackers, scholars, and activists like Niels ten Oever, Joana Varon (executive directrix of codingrights.org), Corinne Cath, and others have worked for years to develop the IETF guidelines for Human Rights Protocol Considerations. This document translates human rights concepts into technical terms relevant for those working on internet networking protocols.[52] It also builds on existing standards that were developed to explicitly support privacy at the network protocol level.[53] Proposed methods include analyzing draft IETF standards for whether they consider human rights at all, analyzing the potential human rights impact of standards changes, incorporating interviews with directly impacted people and communities into the regular process of internet standards design, and post-hoc analysis of the human rights impact of new standards implementations. Guidelines for engineers to consider when developing new protocols include their impacts on privacy, internationalization, open standards, accessibility, authenticity, and anonymity.

Design standards that potentially support social justice are also sometimes adopted into law, as in the case of the universal and accessible design standards that informed the Americans with Disabilities Act in the United States[54] or the General Data Protection Regulation (GDPR) in the European Union.[55] In this book, I have barely touched on law and policy. I hope that the legal and policy implications of design justice will be taken up by legal scholars, public-interest lawyers, and advocacy organizations over time.

A Note about Appropriation

There's a disjuncture between academic attention to community appropriation of technology, as discussed in chapter 3, and the more widely used sense of the term to describe cultural theft. Although many scholars valorize resistant, critical, or bottom-up forms of technological appropriation, appropriation is a process that can be employed by anyone, including those who hold very different values. In popular culture, the term most often is used to name the process whereby white people, and cultural industries that produce and valorize whiteness, constantly steal and use B/I/PoC cultural practices (ideas, fashion, music, food, slang, and so on) without acknowledging their history and origins and without sharing the benefits (monetary and otherwise) that accrue. Those in positions of power under white supremacy benefit from the systematic appropriation, or theft, of ideas and culture from B/I/PoC. Settlers benefit from the appropriation (theft) of native lands and cultures. Under capitalism, the dynamic of appropriation by those in positions of structural power also can be seen in labor process innovations by workers that result in their own displacement through automation. When shop floor workers redesign assembly processes to be more efficient, for example, the gains are typically realized by factory owners, rather than workers, whose workload doesn't decrease; on the contrary, they are frequently expected to produce at a higher level within the new, redesigned process. Under heteropatriarchy, men do not systematically appropriate femme styles, mannerisms, speech forms, and culture; indeed, to do so is to break male gender norms and invite transmisogynistic violence. However, men constantly appropriate labor by women and femmes, including emotional, affective, and reproductive labor, as well as housework and other forms of work that are feminized, racialized, and devalued under heteropatriarchal racial capitalism.[56]

Although many individual designers and developers do not *intentionally* participate in theft from and exploitation of marginalized communities, they do indeed participate in such processes; this is why one of the principles of design justice is that it focuses on outcomes over intent. In many cases, unintentional appropriation plays a key role in reproducing the matrix of domination.

Practices

Chapter 2 explored the question, "Who participates in and controls design processes?" It also argued for accountability to marginalized communities and, ultimately for community control of design. Frequent critiques of community-controlled design processes include variants of "design by committee produces mediocrity" or "we don't want to end up with lowest-common-denominator design!"

"Design by Committee Produces Mediocrity"

There are various versions of the argument that design justice in practice produces mediocre outputs. Among software development communities, for example, the phrase *design by committee* is often shorthand for a process that is assumed to produce designs that are "(a) ineffective, (b) inelegant and (c) not responsive to the core concerns."[57] The implication is that shared decision making never works. To take an example from another domain, in the documentary film world, many directors feel that community accountability crushes creativity.[58] These kinds of arguments must be situated within a larger conversation about the relationship between community accountability, democratic processes, shared decision making, and delegation, on the one hand, and the role of expert knowledge, professionalization, and individual creativity on the other. To address this line of critique, at least two questions are helpful.

First, does design justice require design by committee? The answer is simple: it does not. On the contrary, in a well-functioning design process, the design team recognizes and values the unique skillsets and experiences of each participant. The team frequently delegates particular kinds of work and particular kinds of decisions to skilled individuals and working groups. For example, if one person on the team is a skilled illustrator, they may be assigned the task of creating illustrations and detailed mock-ups for the project. There is nothing about design justice as a framework that necessarily implies that particular talents or skills must be devalued or subordinated to an abstract "collective will." Indeed, if anything, design justice ensures that all of those who contribute to a design process receive recognition, attribution, and, where appropriate, remuneration for their labor. This is in contrast to other design approaches where those at the top of the hierarchy receive the vast bulk of the rewards for the collective labor and ideas of those below them in the pyramid. In contrast, in HCD and even in many participatory design processes, community members who take part in various stages of design, and whose ideas and feedback may provide the key to the realized product, are rarely compensated or recognized. If they are, such recognition is typically token.

Second, does design by committee always produce mediocrity? Perhaps not. To take the most visible example, no one disputes that the internet itself was designed by community consensus.[59] It is fair to say, though, that the devil is in the details. A design process where every decision is made by many people may take much longer, and it is also possible that the results may be mediocre. However, this has more to do with the specific decision-making process of the design committee than with the mere fact that the decision involves a committee at all. Is it consensus? Majority rules? Instant runoff voting? Is it consultative, with a delegated individual making the final decision after listening carefully to input from everyone? For example, standards for the World Wide Web are set by a technical body called the W3C, which functions by committee. The W3C recommends committees of about ten to fifteen people, with a small, highly engaged core augmented by feedback from a larger public mailing list.[60] Software developer and scholar of consensus process Charlie DeTar wrote an excellent doctoral dissertation about the design of sociotechnical systems to support democratic decision making.[61]

Funnel or Prism? A Further Response to the Lowest Common Denominator

The concept of *lowest-common-denominator design* holds that when many people are involved in a design decision, they may arrive at a solution that no one really loves but that everyone can live with. The argument is that design justice asks us to design for everyone, but if we try to design for everyone, we will design boring, uninspired objects. Further, we won't be able to take advantage of all the possible affordances of designed objects if we're trying to make them accessible to all. For example, if we want to design physical spaces that are accessible to people in wheelchairs, then we won't be able to use stairs as a design element in the built environment.

I am writing these words in a small, beautiful house in Punta del Diablo, Uruguay. The house has one large room with high ceilings, a smaller bedroom, and a bathroom. It is designed with a loft area with two child-sized beds; the loft area is accessed by a steep wooden stairway with a ten-step ladder built against the wall at an angle of about 75 degrees. The loft area is clearly inaccessible to anyone who cannot climb such a ladder; the design excludes small children, elders, and many Disabled people. Does design justice imply that we should never build such a loft space? It does not. The loft is an excellent use of space. It takes a small-footprint floorplan and adds an aesthetically pleasing, functional, additional sleeping and working space. It provides an area slightly separate from the bedroom and the main living space. From the standpoint of anyone who enjoys lofts, it is a lovely design decision.

Design justice doesn't imply that we must somehow reduce our options to only those that satisfy all accessibility criteria for the most marginalized within the matrix of domination. It is not meant to be a filter that we use to eliminate most design possibilities from consideration because they fail an accessibility checklist. In fact, design justice as a framework is meant to do the opposite: to act not as a funnel that we use to limit ourselves to a minimal set of supposedly universal design choices, but rather as a prism through which to generate a far wider rainbow of possible choices, each better tailored to reflect the needs of a specific group of people.

As discussed throughout this book, many design approaches attempt to universalize, without acknowledgment of who will benefit, who will be excluded, and who might be harmed. Design justice makes these choices explicit. It is opposed to false universalization; it is allied with standpoint theory. It is an approach to design that recognizes, respects, and specifies difference, instead of pretending to erase difference. Design justice builds on feminist epistemologies.[62] This means that instead of pretending to design based on supposedly universal, unemotional, and value-free data (often a mask for the lived experience of relatively wealthy white cis men), design justice values insights that are developed through open dialogue, empathy, and the lived experience of people from the communities that will be the most affected by the designed object or system, as well as by the design process itself. To return to the example of the small house in Uruguay: the loft, with its ladder steps, is inaccessible to many but is still a wonderful feature of the house. It provides great joy to a certain subset of people, although others cannot make use of it. On the other hand, the narrow width of the only door, together with the six-inch raised lip of the doorway, greatly reduce the accessibility of the entire house to those who use wheelchairs. A design justice approach might indeed support a different design, with wider doors flush at the entrance, to allow an entire family to enjoy time here together, including elders and/or others who might need the use of a wheelchair or who have a harder time with steps.

Design justice, in other words, requires that we specify, consider, and intentionally decide how to best allocate both benefits and harms of the objects and systems we design, with attention to their use context. It doesn't mean lowest-common-denominator design. Quite the opposite: it means highly specific, intentional, custom design that takes multiple standpoints into account. It is not about eliminating the benefits of excellent design unless everyone can access them; instead, it is about more fairly allocating those benefits.

Narratives

Chapter 3 asked, "How do the stories that we tell shape design?" It argued that narrative shifts are necessary in terms of design framing, scoping, and attribution.

Design Saviors versus Design's Role in the Cycle of Struggles

One necessary narrative shift would turn us away from technochauvinism[63] and solutionism and toward an understanding that designers can play an important role within broader social movements. The explicit politicization of design is periodic: it rises and falls within the context of cycles of struggle.[64] As I write these words, we are living through the ascendance to political power of hard-right and explicitly white supremacist tendencies within democracies around the world, from the Trump administration in the United States to Bolsonaro in Brazil. Yet resurgent authoritarianism, like the Snowden revelations, spiraling income inequality, climate crisis, the perpetual War on Terror, and the ever-expanding prison industrial complex, also provoke new social movements, from Occupy Wall Street to #BlackLivesMatter. The continued push for petroleum extraction, linked with ongoing projects of settler colonialism, faces resistance from a new wave of indigenous-led organizing, such as #StandWithStandingRock. The extreme and open racism, misogyny, and xenophobia of the Trump administration galvanized a massive cycle of struggles, including for immigrant rights (#NoWallNoBan and #KeepFamiliesTogether), against rape culture (#MeToo), and more. These take place at the same time as the increased mainstream cultural visibility of trans* people of color, the spread of intersectional analysis, and the repoliticization of queer struggle. It's in this context that many feel a desire to realign design values, practices, narratives, sites, and pedagogies with explicitly intersectional feminist, queer, antiracist politics. Design justice thus is part of a broader cycle of struggles.

Platform Cooperativism versus the "Sharing Economy"

Another key narrative that must be challenged by design justice practitioners is that of the so-called sharing economy. The design of platforms like Uber, Amazon, Airbnb, and other digital markets for on-demand services and goods is currently structured to reinforce the power of consumers over workers, and owners over all. Platform affordances too often privilege those who occupy positions of social and economic privilege. Platform design is a key "moment" in the reproduction of economic relationships and social control under white supremacist capitalist heteropatriarchy and settler colonialism. Platform ownership also is an increasingly important source of capitalist profitability and worker exploitation. Counterstrategies to the "Uberization of everything" include worker self-organization, consumer boycotts and buycotts, shareholder activism, platform worker organizing by labor unions, and platform cooperativism.

Platform cooperativism is the proposal, most clearly articulated by media studies scholar-activist Trebor Scholz, journalist and media studies professor Nathan Schneider, and lawyer and writer Janelle Orsi, that workers should own their own digital labor markets.[65] There is a growing volume of writing on platform cooperativism, as well as a community of practice that has formed around the conference of the same name and the Platform Cooperativism Consortium (platform.coop.net). Examples of already existing platform cooperatives include photographer-owned stock photography platform Stocksy, musician- and listener-owned streaming service resonate.is, and driver-owned Green Taxi Co-op in Denver. Other platforms that aren't cooperatives but have been designed together with workers to support worker power include Contratados.org (a "Yelp for migrant workers" by the Center for Migrant Rights), Turkopticon (where Mechanical Turk workers can share resources and information about employers), Alia (a portable benefits platform for home cleaners by the National Domestic Workers Alliance), and many others. Platform cooperativism is an important proposal with a growing group of adherents. At the same time, platform cooperativism will not be able to advance as a liberatory project if it fails to fully incorporate race and gender analysis, and it will advance most fruitfully if its practitioners integrate a design justice approach. Design justice, applied to the development of digital labor markets, means involving workers, worker advocacy organizations, and cooperatives from the beginning in the design of (cooperative, worker-owned) platforms in various sectors.

There are many other master narratives about design that must be challenged and replaced; unpacking them is one of many important tasks for design justice practitioners in the future.

Sites

Chapter 4 asked, "How do we imagine and construct more intentionally liberatory sites where design justice principles can come to life?" Privileged design sites are raced, classed, and gendered. We need to challenge the ways that the matrix of domination is reproduced within design sites like hackathons, hacklabs, makerspaces, and fablabs. Hopefully, these can be transformed into deeply diverse and inclusive spaces, and chapter 4 documents many ways that this is already happening. At the same time, we need to think beyond diverse participation alone, to consider how such sites might be reconfigured to help hard-code liberation, shift discursive power, and instantiate design justice pedagogies.

Although the chapter concludes with some suggestions for how to make design sites more inclusive, practical guides to organizing various kinds of sites according to design justice principles (like the *DiscoTech* zine) would be useful. Victoria Palacios has recently synthesized a set of extremely helpful guidelines from existing literature and how-to guides; her work is available at bit.ly/designeventguidelines. Besides opening privileged design sites to more people—in particular, those who are marginalized and multiply burdened under the matrix of domination—we also need to valorize and systematically resource subaltern design sites. In addition to actions that individual space or event organizers can take, we also need to think on the policy level. For example, what would it look like for cities, states, and countries to condition permitting, site allocation, and grants to hacker, maker, and innovation spaces in part based on diversity and inclusion plans and measurable targets?

In addition, although chapter 4 is an attempt to think about design sites through a design justice lens, it is beyond my capacity to elaborate a spatial theory of design justice, to deeply engage with the extensive literature in architecture and urban planning, or to do justice to the many people and organizations who already do that work. For example, the

Design Justice Platform, initially convened by architect Bryan C. Lee Jr. and his firm Colloqate Design, organized a series of local DesignAsProtest events in cities including New Orleans, New York, and Detroit on January 20, 2017. These events gathered architects and city planners to act in solidarity with and defend the communities most targeted by the incoming Trump administration. In September of 2018, the same group organized a Design Justice Summit in New Orleans, in affiliation with the American Institute of Architects (AIA).[66] The EquityXDesign group has developed an analysis of gender and racial disparity in architecture, organized a series of conferences, pushed the AIA to collect data and set targets for equity, conducted a series of surveys of professional architects and produced publications about ongoing disparities in the field, and created public-facing campaigns to demand equity in architecture.[67] In the future, it will be important for the Design Justice Network to develop closer ties with groups focused on design justice within architecture, urban planning, and related design domains.

Pedagogies

Chapter 5 focused on the question, "How do we teach and learn design justice?" It built on popular education methods and explored design justice pedagogies in both formal and informal educational spaces. In the chapter, I draw largely from my own experience teaching in a university setting. Some questions for further exploration include: What would it mean for institutional structures to support community-engaged pedagogies of technology design? And what are the challenges to realizing design justice pedagogies in an age of the neoliberalization of the educational system?

For example, the Boston Civic Media Consortium links educators from universities across the Greater Boston area who work with PAR, PD, or codesign approaches. In 2018, the consortium released a report that summarizes some of the key challenges to this kind of engaged pedagogy.[68] In the realm of institutional support, there is also a recent boom in tech ethics classes. This is driven in part by the public conversation about ethics and AI, as well as by funders like Omidyar, Mozilla, Schmidt Futures, and Craig Newmark Philanthropies, which in 2018 partnered to launch the Responsible Computer Science Challenge. This grant competition supports the creation of classes that integrate ethics into undergraduate CS training.[69]

At the same time, future work should explore design justice pedagogies in other learning sites. For example, there is an urgent need to discuss how to teach and learn design justice specifically with younger children, in high schools, and in community colleges, as well as to unpack the relationship between design justice approaches and the numerous different kinds of coding boot camps. For example, the design studio And Also Too has recently launched a *Consentful Tech UX/UI Un-Bootcamp*, which they describe as "a net-for-profit education program that will equip learners with digital design skills and result in prototypes of consentful tech."

Simultaneously, there is a need to interrogate the dynamics that lie behind recent attempts to apply design thinking to education, and to teach all students design thinking. In a brilliant summary of growing pushback against calls to "rethink education," educator Sherri Spelic writes: "Design Thinking aligns well with a certain kind of neoliberal enthusiasm for entrepreneurship and start-up culture. I question how well it lends itself to addressing social dilemmas fueled by historic inequality and stratification."[70] As she notes, any approach to redesigning education that leaves history and structural inequality out of the picture is not an approach that will turn out well for youth of color, low-income youth, and others who have always been marginalized by the formal educational system. Instead, she argues: "Our students can see inequality. Many of them experience its injustices on a daily basis. Precisely here is where I would like to see us focus our educator energies: on helping students see and identify the faulty designs throughout our society that plague the most vulnerable among us. In order to dismantle and correct these designs and patterns, they must first be able to notice and name them. That's the kind of design thinking I hope and wish for: Where 'what's wrong?' drives our pursuit of 'what if?'"[71] Finally, the discussion of design justice pedagogies must be more closely linked with the movement for educational justice. This means connecting design justice work with student, teacher, and parent-led community organizing groups that focus on education, like Philly Student Union, People in Education in Detroit, Make the Road New York, Youth Justice Coalition in Los Angeles, and many others. It also means linking with national networks like the Alliance for Educational Justice and with the new wave of teacher union organizing, such as the successful United Teachers Los Angeles (UTLA) strike. These and many other groups have been fighting for years to end the war on youth of color, dismantle the school-to-prison pipeline, and build power among youth organizers to demand quality education for all. True design justice pedagogies will be more tightly connected to youth-, teacher-, and parent-led struggles around the future of education.

Conclusions

#TechWontBuildIt is an exciting development. Successful worker-led campaigns to push Google to abandon Project Maven, to cancel Project Dragonfly, and to take #MeToo seriously are all important, as are the ongoing campaigns at Microsoft, Amazon, IBM, Salesforce, and other tech companies to end complicity with ICE's ongoing human rights violations. What's more, #TechWontBuildIt mobilizations are not single-issue campaigns. Many of the workers involved have built coalitions across firms and are tightly linked to the networked social movements that characterize the present cycle of struggles. It remains to be seen whether the current mobilization wave is a short-lived moment in response to the polarized political climate under the Trump administration, or the beginning of a sea change that has the potential to reshape the trajectory of sociotechnical design writ large.

Either way, hopefully design justice as an approach, and the growing Design Justice Network, can help provide some useful concepts and tools. Although design practices today too often systematically reproduce the matrix of domination, there is a growing community of design justice practitioners: people and organizations who work on a daily basis to leverage the power of design for collective liberation and ecological sustainability. I hope that this book has provided a window into that work. Together, let's build the worlds we need!

Glossary

This glossary includes acronyms, as well as short descriptions of some of the key terms used throughout the book. Key term definitions are mostly based on those provided by the Transformative Media Organizing Project at transformativemedia.cc/research. *Glossary compiled by Annis Rachel Sands.*

18F. A federal office tasked with supporting other government agencies to build and improve tech products and services.

+KAOS. A collectively authored history of the Italian radical tech collectives Autistici/Inventati.

#MeToo. Social movement started by Tarana Burke in 2006 using the then-popular social media website Myspace to bring visibility to the ongoing sexual assault and harassment experienced by Black women. In 2017, actress Alyssa Milano adopted the hashtag #MeToo to challenge the sexual violence and harassment experienced by women and femmes in Hollywood. Initially Milano was credited with starting #MeToo, despite Burke's decade-long use of #MeToo. However, Milano and the Time's Up movement credited Burke and the work done by other Black and brown women activists and organizers to draw visibility to the most marginalized women and femmes around the world who experience gender and sexual violence. See **Time's Up, #TIMESUP**.

#MoreThanCode. A participatory action research project about the field of technology for social justice. Explore the *#MoreThanCode* report at https://morethancode.cc.

#TIMESUP. The hashtag for the Time's Up campaign that emerged from the #MeToo movement. The campaign officially launched on January 1, 2018. See **Time's Up**.

A/B testing. Randomized experiments to compare and test the performance of two different variants, often used in web design.

A/I. Autistici/Inventati. Italian hacker activist collectives, authors of the book +Kaos.

ACLU. American Civil Liberties Union.

ACT UP!. AIDS Coalition to Unleash Power.

ADA. Americans with Disabilities Act.

AI. Artificial Intelligence.

AIDS. Acquired Immune Deficiency Syndrome.

AJN. *American Journal of Nursing.*

AMC. Allied Media Conference.

AMP. Allied Media Projects.

AORTA. A worker-owned cooperative devoted to strengthening movements for social justice and a solidarity economy. Explore http://aorta.coop.

API. Application programming interface.

ARRA. American Recovery and Reinvestment Act, more commonly known as the Obama Stimulus Bill.

B/I/PoC. Black/Indigenous/people of color.

BBS. Bulletin board system.

BTOP. Broadband Technology Opportunity Program.

C-Innova. *Centros de Innovacion Comunitaria*, or community innovation centers in English.

CBOs. Community-based organizations.

CCTV. Cambridge Community Television. CCTV is the new name for the organization formerly known as NeighborMedia. See **NeighborMedia**.

CERO. Cooperative Energy, Recycling, and Organics, a cooperatively owned commercial composting company based in Dorchester, Massachusetts. Explore http://www.cero.coop.

CIL. Civic Innovation Lab.

Cis. Short for cisgender. See **Cisgender**.

Cisgender. Nontransgender. Someone whose gender identity is consistent with the sex they were assigned at birth. For example, a person who is assigned male at birth, is seen by others as male, and whose gender identity is male is a cisgender male (cis male). Also shortened to cis, as in cis man, cis woman.

CL/VU. City Life/Vida Urbana. A Boston-area housing rights organization and anchor member of the national Right to the City alliance.

Co-op. Co-operative.

CRMs. Constituent relationship management systems.

CS. Computer science.

CUTgroup. Chicago User Testing group.

DARPA. Defense Advanced Research Projects Agency.

DC. District of Colombia.

DCTP. Detroit Community Technology Project.

DDJC. Detroit Digital Justice Coalition.

DHS. Department of Homeland Security.

Disabled people. While some prefer to use the "people-first" term "People with Disabilities" (often abbreviated PwD), I use "Disabled people," or "identity-first" language throughout this book. While both terms emerged from disability advocacy, some feel that the former implies the individual/medical model of disability, while the latter is tied more closely to disability justice conceptions of the social production of disability. For a brilliant recent work on disability justice see Piepzna-Samarasinha 2018.

DiscoTechs. Discovering Technology events, originally created by the Detroit Digital Justice Coalition.

DIT. Do-it-together.

DIY. Do-it-yourself.

DMV. Department of Motor Vehicles.

DREAM. Development, Relief, and Education for Alien Minors Act.

DS4SI. Design Studio for Social Intervention.

EBIT. Earnings before Interest and Taxes.

ENIAC. Electronic Numerical Integrator and Computer. According to computerhistory.org, in 1942 physicist John Mauchly proposed the need for an "all-electronic calculating machine." For two years, between 1943 and 1945, the US Army invested resources, time, and personnel to develop this vision, leading to the creation of ENIAC, "the first large-scale computer to run at electronic speed without being slowed down by any mechanical parts."

F/LOSS. Free/libre and open-source software.

Fablab. Fabrication laboratory.

FAT*. Fairness, Accountability, and Transparency. A conference about bias in machine learning, natural language processing, AI, and other computing processes.

FTAA. Free Trade Area of the Americas.

FTAA IMC. FTAA Independent Media Center.

GIFs. Graphics Interchange Format.

GNC. Gender non-conforming. Can be taken to include identities that are genderqueer, gender variant, gender fluid, or third gender, as well as those that are bigendered, multigendered, nonbinary, nongendered, androgynous, masculine-of-center, feminine-of-center, and gender-questioning, among other gender identities.

GSA. General Services Administration.

HCD. Human-centered design.

HCI. Human-computer interaction.

IBM. International Business Machines Corporation.

ICE. US Immigration and Customs Enforcement.

ICTs. Information and communication technologies.

IDDS. International Development Design Summits, or *Cumbres Internacionales de Diseño para el Desarollo.*

IDEO. A global design and consulting firm founded in 1991.

IDEPSCA. Institute of Popular Education of Southern California.

IDRC. Inclusive Design Research Centre.

IMC. Independent Media Center. Also called **Indymedia.**

Indymedia. Alternative name for Independent Media Center. See **IMC.**

Intersectionality. Intersectionality (following feminist legal scholar Kimberlé Crenshaw) refers to the ways that structural oppression is not based only on race or gender identity, but on the intersection of race, gender identity, sexual orientation, class, immigration status, disability, age, and other axes of identity.

Intersex. Intersex individuals are persons who, at birth, cannot be classified according to the medical norms of male and female bodies with regard to their chromosomal, gonadal, and/or anatomical sex. The term *inter** has also been used as an umbrella term that denotes the diversity of intersex realities and bodies.

IPVtech. Intimate partner violence and technology.

IRC. Internet Relay Chat.

Jot@. The non-gender-specific form of the Spanish language term *Joto* or *Jota,* a once-derogatory Mexican slang term that has been reappropriated as a term of community pride, similar to the word *Queer.*

K–12. Kindergarten through twelfth-grade education.

LA. Los Angeles.

LAPD. Los Angeles Police Department.

Latinx. Gender-neutral term for people who identify as being Latin American or having ancestral and/or family roots from Latin America.

LGBTQ. Lesbian, gay, bisexual, trans*, queer.

LGBTQI. Lesbian, gay, bisexual, trans*, queer, intersex.

LGBTQIATS. Lesbian, gay, bisexual, trans*, queer, intersex, asexual, two-spirit.

LLK. Lifelong Kindergarten. LLK is a research group in the MIT Media Lab.

Loconomics. A freelance jobs platform that is like a cooperatively owned version of TaskRabbit. See https://www.loconomics.coop.

LOL. Liberating Ourselves Locally. See **LOLspace**.

LOLspace. A queer and trans* people of color–centered, social justice–focused makerspace in East Oakland. Launched in 2011, merged with Peacock Rebellion in 2017, and led by a crew of hackers, healers, artists, and activists who are queer and trans* people of color. See **LOL**.

M-Pesa. Mobile money transfer service launched by Vodaphone in 2007.

Matrix of domination. Black feminist sociologist Patricia Hill Collins developed this term to refer to the linked systems of white supremacy, patriarchy, capitalism, and settler colonialism.

Media justice. Media justice is "a long-term vision to democratize the economy, government, and society through policies and practices that ensure: democratic media ownership, fundamental communication rights, universal media and technology access, and meaningful, accurate representation within news and popular culture for everyone." Explore MediaJustice.org.

MIT. Massachusetts Institute of Technology.

MOOCs. Massive open online courses.

MOU. Memorandum of understanding.

NASA. The National Aeronautics and Space Administration.

NCWIT. National Center for Women and Information Technology.

NDLON. National Day Labor Organizing Network.

NDWA. National Domestic Workers Alliance.

NeighborMedia. Former name for CCTV. See **CCTV**.

NGO. Nongovernmental organization.

NIDP. Norwegian Industrial Democracy Project.

NPR. National Public Radio.

NuVu. "NuVu is a full-time innovation school for middle and high school students in Cambridge, MA. NuVu's pedagogy is based on the architectural Studio model and geared around multi-disciplinary, collaborative projects. We teach students how to navigate the messiness of the creative process, from inception to completion by prototyping and testing." Explore https://cambridge.nuvustudio.com/pages/what-is-nuvu.

NYC. New York City.

NYPD. New York Police Department.

OCAD. OCAD Toronto is the present name for the former Ontario College of Art and Design.

Odeo. Podcasting company where Twitter was born. See also **TXTMob**.

OLPC. One Laptop per Child.

Online organizing. Online organizing is often used to describe internet- and mobile-phone-focused approaches to political and social movement campaigns. For example, a typical job description for an online organizer looks for someone who will write and implement email blasts, run SMS/text-messaging campaigns, manage social media accounts, and so on.

OTI. Open Technology Institute.

PAD. Participatory action design.

PAR. Participatory action research.

PARTI. Participatory Artistic Traveling Installation. PARTI raises public awareness about Urbano and engages diverse communities in imagining their emancipated City of Boston. See https://codesign.mit.edu/2013/12/urbano-parti/.

Participatory media. Forms of media that are designed to allow, invite, and encourage many people to take part in the process of media production—in other words, to make their own media or tell their own stories. Participatory media are not always online and can include cultural forms such as art-making, music, and dance, among others.

PD. Participatory design.

PoC. People of color.

Pop ed. Popular education.

PwD. People with disabilities.

QT. Queer, trans*.

QTI/GNC. Queer, trans*, intersex, and/or gender-non-conforming.

QTPOC. Queer, trans*, people of color.

Queer. Queer is used as a broad umbrella term by a wide range of people who identify as outside of normative and/or binary constructions of gender, gender identity, sex, and/or sexual orientation. The term is fluid rather than fixed (it doesn't mean just one thing).

R&D. Research and development.

RAD. Research Action Design, a worker-owned collective that "uses community-led research, collaborative design of technology and media, and secure digital strategies to build the power of grassroots social movements." Explore http://rad.cat.

RTC. Radical tech collectives.

RNC. Republican National Committee.

School push-out. Discriminatory disciplinary practices, among other factors, produce elementary, junior high, and high school noncompletion rates that are much higher among youth of color than among white youth, among LGBTQ youth than among straight youth, and highest among LGBTQ youth of color. *Push-out* rather than *drop-out* emphasizes the structural, systemic, and institutional forces beyond young people's control that contribute greatly to school noncompletion.

SIGCSE. Special Interest Group on Computer Science Education.

SMS. Short message service.

SOGI. Sexual orientation and/or gender identity.

SoMove. The Social Movements Oral History Tour.

SpideyApp. An Android-based Stingray detector.

Stanford d.school. Hasso Plattner Institute of Design at Stanford University.

STEM. Science, technology, engineering, mathematics.

STS. Science and technology studies.

T4SJ. Tech for Social Justice Project, a participatory action research project that produced the *#MoreThanCode* report. Explore https://morethancode.cc.

TecnoX. A growing network of open hardware hackers from across Latin America who are engaged in conversations about how to connect open hardware hacking to social movements. Explore tecnox.org.

Time's Up movement. A movement founded by women and femmes in Hollywood against sexual harassment. Since its launch on January 1, 2018, the movement has raised over $22 million for its legal defense fund. See **#MeToo** and **#TIMESUP**.

Trans*. This book uses trans* to broadly include people whose gender identity differs from the gender they were assigned at birth. Trans* may include (among other identities and communities) transgender, transfeminine, transmasculine, MTF (male-to-female), FTM (female-to-male), genderqueer, gender-non-conforming, gender-variant and third gender/sex, transsexual, two-spirit, and transvestite/cross-dresser.

Trans*H4CK. A trans* hackathon, speaker series, and code school. Explore transhack.org.

Transformative media organizing. "Transformative media organizing is a liberatory approach to integrating media, communications, and cultural work into movement building. It lies at the place where media justice and transformative organizing overlap. Transformative media organizers begin with an intersectional analysis of linked systems of race, class, gender, sexuality, ability, and other axes of identity. We seek to do media work that develops the critical consciousness and leadership of those who take part in the media-making process, create media in ways that are deeply accountable to the movement base, invite our communities to participate in media production, create media strategically across platforms, and root our work in community action." For more information, explore http://transformativemedia.cc.

TSA. Transportation Security Administration.

Two-Spirit. Among some Indigenous North American cultures, Two-Spirit refers to individuals whose spirits are a blending of male and female. Two-Spirit is essentially an umbrella term for third genders recognized in many Indigenous cultures. For more information, explore the Northeast Two-Spirit Society.

TWTTR. The original project code name for what is now known as Twitter.

TXTMob. An experimental group SMS application that was developed by Tad Hirsch, who at the time was a graduate student at the MIT Media Lab. Inspired **TWTTR**.

UCD. User-centered design.

UCIC. Upham's Corner Input Collector. An interactive public planning installation by DS4SI and students from the MIT Codesign Studio.

UCLA. University of California, Los Angeles.

UD. Universal design.

UI. User interface.

UK. United Kingdom.

USA. United States of America.

USB. Universal Serial Bus.

USC. University of Southern California.

Userforge. Allows rapid random generation of user personas. Explore Userforge.com.

USSR. Union of Soviet Socialist Republics.

UTOPIA project. The canonical first successful instance of participatory design (see **PD**). UTOPIA was a collaboration among the Nordic Graphic Workers Union, researchers, and technologists, who worked with newspaper typographers to develop a new layout application.

UX. User experience.

UYC. Urban Youth Collaborative, a New York City–based youth organizing group.

UYC SMS Survey Initiative. An SMS survey system to gather data about students and their experiences of surveillance and police abuse inside New York City high schools, in collaboration with UYC. See **UYC**.

VC. Venture capital.

VCs. Venture capitalists.

VSD. Value-sensitive design.

WTO. World Trade Organization.

ZUMIX. A youth music and media organization in East Boston. Explore https://www.zumix.org/about/history.

Notes

Preface

1. Critical Art Ensemble 2008; Costanza-Chock 2014; Dizikes 2014; Jenkins et al. 2016.

Introduction

1. See alliedmedia.org.

2. For a recent discussion of the increasingly widespread use of the term trans* with an asterisk, see Halberstam 2018.

3. Sadat 2005.

4. Schneier 2006.

5. Despite my participation in social movement networks, including the global justice movement, Indymedia, the immigrant rights movement, countersurveillance work, and more, my white skin, institutional affiliations, educational background, and US citizenship have largely protected me from the most egregious types of abuse by state power.

6. Costello 2016.

7. Irani 2016; Dyer-Witheford 2016; and Gray and Suri 2019.

8. See https://www.tsa.gov/transgender-passengers.

9. Winner 1980.

10. As Anna Lauren Hoffmann notes about the simplified gender binary interface, "The thing that really gets me is that this screen was developed as a privacy-preserving compromise after folks realized the level of detail these machines were *actually* capable of rendering!" Twitter, September 3, 2018, https://twitter.com/annaeveryday/status/1036635912761819136.

11. In 2009, Toby Beauchamp wrote about state surveillance and trans* concealment/visibility (Beauchamp 2009). In September of 2016, Shadi Petosky brought national attention to the challenges of #TravelingWhileTrans when she live-tweeted her experience with an invasive search by TSA agents at the Orlando airport, after she was flagged in a millimeter wave scan for presenting as female while having a penis. See Lee 2016.

12. See https://www.propublica.org/article/tsa-not-discriminating-against-black-women-but-their-body-scanners-might-be.

13. Browne 2015.

14. Buolamwini 2017.

15. Throughout this book I use the identity-first term "Disabled people" rather than the people-first term "people with disabilities" because design justice is more closely aligned with a social/relational disability justice analysis than with the individual/medical model of disability. For more, explore Piepzna-Samarasinha 2018.

16. Ito 2017.

17. The seeds for this gathering were planted in 2015 at the Future Design Lab at AMC, itself inspired by the Discovering Technology events, or DiscoTechs, organized by the Detroit Digital Justice Coalition. See https://www.alliedmedia.org/ddjc/discotech.

18. The authors of the first version of the Design Justice Network Principles are Una Lee, Jenny Lee, Melissa Moore, Wesley Taylor, Shauen Pearce, Ginger Brooks Takahashi, Ebony Dumas, Heather Posten, Kristyn Sonnenberg, Sam Holleran, Ryan Hayes, Dan Herrle, Dawn Walker, Tina Hanaé Miller, Nikki Roach, Aylwin Lo, Noelle Barber, Kiwi Illafonte, Devon De Lená, Ash Arder, Brooke Toczylowski, Kristina Miller, Nancy Meza, Becca Budde, Marina Csomor, Paige Reitz, Leslie Stem, Walter Wilson, Gina Reichert, and Danny Spitzberg. The designjusticenetwork.org website includes blog posts that further describe the origins of the network; for example, to learn more about the first workshop at AMC, see http://designjusticenetwork.org/blog/2016/generating-shared-principles.

19. The Design Justice Network Principles and list of signatories are available at http://designjusticenetwork.org/network-principles.

20. The Design Justice Network has been built through the hard work of many, many people over the past several years. It would be difficult to list every individual, group, and community here. Many additional track coordinators are named in the acknowledgments at the beginning of this book, and can be found in the Allied Media Conference program books and on the Design Justice Network website.

21. See https://www.alliedmedia.org/amc/previous-years.

22. See https://www.andalsotoo.net.

23. And Also Too is known for projects such as graphic design with the Feathers of Hope First Nations Youth Action Plan; an infant feeding resource with HIV positive mothers with CATIE, the Teresa Group, and Women's College Hospital; and Contratados.org, a resource for migrant workers, with Research Action Design, Studio REV-, and the Centro de los Derechos del Migrante; among many other projects.

24. I was a co-founder of RAD.

25. See EquityXDesign 2016.

26. See https://idrc.ocadu.ca/about-the-idrc.

27. For more information explore https://www.civicdesigner.com, see also McDowell and Chinchilla 2016.

28. Chardronnet 2015.

29. For a list of people and organizations who have signed the Design Justice Principles, see http://designjusticenetwork.org. For expanded lists of organizations, networks, and projects working in this space, see https://morethancode.cc and also https://www.ruhabenjamin.com/resources.

30. See Harding 2004 for an edited volume that brings together key scholars of standpoint theory including Dorothy Smith, Donna Haraway, Patricia Hill Collins, Nancy Hartsock and Hilary Rose.

31. Jobin-Leeds and AgitArte 2016; see also https://agitarte.org.

32. Downing 2003; Halleck 2003; and Kidd 2013.

33. See https://archive.org/search.php?query=indymedia.

34. See https://www.alliedmedia.org.

35. VozMob Project 2011.

36. Lewin 1946; Dewey 1933; Freire 1972; Fals-Borda 1987; and Smith 2013.

37. Costanza-Chock et al. 2018; see also https://morethancode.cc.

38. For the *Oxford English Dictionary* definition, see https://www.lexico.com/en/definition/design. See also the *Merriam-Webster Dictionary*: "Design (transitive verb): to create, fashion, execute, or construct according to plan," https://www.merriam-webster.com/dictionary/design.

39. Hoffman, Roesler, and Moon 2004.

40. Furniture designer Charles Eames said that design is "a plan for arranging elements in such a way as to best accomplish a particular purpose." Quoted in Neuhart et al. 1989.

41. In the original, Papanek says, "All men are designers."

42. Papanek 1974, 17.

43. In his 2010 book, *Design as Politics,* Fry also specifies at least three separate meanings of *design* that are often conflated: first, the design object; second, the design process; and third, the design agent, which may be an individual designer, a design firm, or an array of people and sociotechnical processes (what Latour might call an *actor-network*) that engages in design activities. See Fry 2010.

44. Willis 2006, 80.

45. Dalla Costa and James 1972.

46. Shetterly 2017.

47. Von Hippel 2005.

48. See https://www.aiga.org.

49. See https://www.access-board.gov.

50. Hoffman, Roesler, and Moon 2004, 89.

51. Hoffman, Roesler, and Moon 2004, 89.

52. Aliseda 2006.

53. Schön 1987.

54. DiSalvo and Lukens 2009.

55. Escobar 2018, 21.

56. Srinivasan 2017.

57. Hernández-Ramírez 2018.

58. Irani 2018.

59. See Natasha Jen's talk at https://99u.adobe.com/videos/55967/natasha-jen-design-thinking-is-bullshit.

60. Benjamin 2019a.

61. Truth 1995 (originally published in 1851); Jones, cited in Davies 2007; Combahee River Collective 1983 (originally published in 1977).

62. Crenshaw 1989.

63. Crenshaw 1989, 144.

64. Crenshaw 1989, 149.

65. Crenshaw 1991.

66. In that article, Crenshaw also goes on to describe structural, political, and representational intersectionality.

67. Crenshaw 1989, 140.

68. Buolamwini and Gebru 2018.

69. Collins 2002.

70. Collins 2002, 229.

71. Collins 2002, 223.

72. Angwin and Grassegger, 2017.

73. Gillespie 2018.

74. Harwell and Miroff 2018.

75. Segarra and Johnson 2017.

76. Collins 2002, 234.

77. Gibson 1979.

78. Friedman 1997.

79. Wajcman 2010.

80. Charlton 1998.

81. Von Hippel 2005; Schuler and Namioka 1993; and Bardzell 2010.

82. Siles 2013.

83. Downing 2000.

84. Maxigas 2012.

85. Irani 2015.

86. See https://codesign.mit.edu.

87. Allied Media Projects n.d.

1 Design Values

1. See https://logicmag.io/03-dont-be-evil.

2. Tweet by Imani Gandy (@AngryBlackLady), September 29, 2014, https://twitter.com/AngryBlackLady/status/516604901883797505.

3. Trans and Queer Liberation + Immigrant Solidarity Protest announcement; see https://www.facebook.com/events/392408787772958.

4. Tufekci 2017.

5. Gerbaudo 2012.

6. Srinivasan 2017.

7. Bailey, Foucalt Welles, and Jackson 2019.

8. Treré 2012; Cammaerts 2015; and Renzi 2015. See also work by Veronica Barassi (2013) and Robert Gehl (2015), among others.

9. Gerbaudo 2012; Adamoli 2012, 1888.

10. For example, see Salsa Labs (https://www.salsalabs.com/blog/practical-steps-engage-supporters); for Arnstein's concept of the "ladder of participation," see Arnstein 1969.

11. De Vogue, Mallonee, and Grinberg 2017.

12. Interaction Design Foundation, "Affordances," 2019, retrieved June 11, 2019, from https://www.interaction-design.org/literature/topics/affordances.

13. Gibson 1979, 127.

14. Gaver 1991.

15. Norman 2006.

16. Norman 2006, 9.

17. Norman 2006, 216.

18. Norman 2006, 6.

19. Norman 2006, 162.

20. See the Union of the Physically Impaired Against Segregation 1975 and Oliver 2013.

21. For a brilliant recent overview of disability justice, see Piepzna-Samarasinha 2018.

22. Norman 2006, 219–220 and 229, respectively.

23. Gaver 1991, 81.

24. Wittkower notes that disaffordances have been little discussed in design literature but traces the concept to Gee and Marcus, who refer to a dissaffordance as a design feature that may "protect from or exclude other species or members of [our] own species" (Wittkower 2016, 4). Future scholarship might further theorize disaffordances in light of Ruha Benjamin's (2019b) discussion of the expanding carceral logic that undergirds the design of technological systems in a growing array of fields, from prisons and criminal justice to housing and health care.

25. See Joy's TED talk about this experience at https://www.ted.com/talks/joy_buolamwini_how_i_m_fighting_bias_in_algorithms.

26. Gaver 1991, 80.

27. Gaver does acknowledge that "whether a handle with particular dimensions will afford grasping depends on the grasper's height, hand size, etc. Similarly, a cat-door affords passage to a cat but not to me, while a doorway may afford passage to me but not somebody taller. Affordances, then, are properties of the world defined with respect to people's interaction with it." Gaver's affordance theory is thus relational, but it fails to acknowledge systematically structured inequality.

28. Wachter-Boettcher 2017.

29. Winner 1980.

30. Winner 1980.

31. Browne 2015.

32. See https://siteselection.com/theEnergyReport/2011/may/sustainable-buildings.cfm.

33. Hurley 2018.

34. Capps 2017.

35. Benjamin 2016a, 147–148. Benjamin also summarizes the concept of antidiscriminatory design in an excellent TEDx talk.

36. Benjamin 2016a, 147.

37. Benjamin 2019a.

38. Benjamin 2019b.

39. Chemaly 2016.

40. Miner et al. 2016.

41. Up to 20 percent of women in the United States will experience rape or sexual assault, while one in four will experience intimate partner violence. See the research digest of the National Coalition Against Domestic Violence in its fact sheet at https://www.speakcdn.com/assets/2497/domestic_violence2.pdf.

42. Chemaly 2016.

43. Mohanty 2013.

44. For a recent book-length discussion of discriminatory design and digital technology, see Wachter-Boettcher 2017.

45. Sue et al. 2007.

46. Tynes, Rose, and Markoe 2013.

47. Gray 2012.

48. Adam et al. 2015.

49. Sue 2010.

50. Accuracy disparities in both sensing and image analysis by skin tone is a recognized and well-documented problem that affects multiple domains. For a recent example, see Buolamwini and Gebru 2018, a widely cited study that has been replicated by several companies, as well as Buolamwini 2017 (disclosure: I was a member of Buolamwini's doctoral committee). See also Coo et al. 2019. However, skin tone disparity in soap dispenser accuracy is a known but understudied problem in peer reviewed research literature. For example, see Rutkin (2016), who summarizes many examples of racial bias in sociotechnical systems and mentions the soap dispenser but does not provide supporting evidence, and Hankerson et al. (2016), who cite a 2015 article by Anupum Pant that provides additional reasoning about why these systems might work better for darker skin tones in India than in the United States, but no empirical evidence. Hankerson et al. also cite a *Slate* article that includes a passage where the reporter states that "Pete DeMarco, the director of compliance engineering at American Standard (the largest toilet manufacturer in the world) ... told me that when automatic fixtures first got popular in the early 1990s, they had difficulty detecting dark colors, which tended to absorb the laser light instead of reflecting it back to the sensor. DeMarco remembers washing his hands in O'Hare Airport next to an African-American gentleman. DeMarco's faucet worked; the black man's didn't. The black guy then went to DeMarco's faucet, which he had just seen working seconds before; it didn't work. This time DeMarco spoke up, telling him to turn his hands palm side up. The faucet worked" (Schulz 2006). See also Benjamin 2019a.

51. Woods 2016.

52. Winner 1980; Latour 1992.

53. Friedman and Nissenbaum 1996, 1997; Friedman, Kahn, and Borning 2002; and Friedman et al. 2013.

54. Friedman and Nissenbaum 1997.

55. Friedman and Nissenbaum 1996.

56. Paul 2016.

57. Muñoz, Smith, and Patil 2016.

58. Benjamin 2019a, 7.

59. Huff and Cooper 1987.

60. For an excellent summary of the literature about design personas and stereotypes, see Turner and Turner 2011. See also Cutting and Hedenborg 2019 for a recent critique of personas.

61. Friedman and Nissenbaum 1997, 39.

62. See Wajcman 2010; Benjamin 2019a, 2019b; and Noble 2018.

63. See https://criticalracedigitalstudies.com/.

64. Kirkham 2015.

65. Bivens 2017.

66. Haimson and Hoffmann 2016.

67. Flanagan, Howe, and Nissenbaum 2008, 327.

68. Friedman and Henry 2019.

69. Williamson 2011.

70. Williamson 2019.

71. Williamson 2019.

72. Hamraie 2017.

73. Alpert 2018.

74. Kafer 2013.

75. Davis 2017.

76. Story 2001.

77. Hamraie 2017.

78. Inclusive Design Research Centre, n.d.; see https://idrc.ocadu.ca.

79. Inclusive Design Research Centre, n.d.

80. Inclusive Design Research Centre, n.d.

81. Kuhn (1962) 1996, 76.

82. Krug 2000.

83. Krug 2000.

84. *Bounce rate* measures the proportion of site visitors who leave the site after viewing only one page. Web developers and site owners want the bounce rate to be as low as possible because that indicates that visitors explore multiple pages on the site (and can be served more advertising).

85. For a regularly updated overview of Spanish use in the United States, see https://en.wikipedia.org/wiki/Spanish_language_in_the_United_States, and for the gold standard data source explore the American Community Survey data at http://data.census.gov.

86. Reinecke and Bernstein 2011.

87. Zuboff 2015.

88. Holmes 2018.

89. Buolamwini and Gebru 2018.

90. For brief overviews, see Caplan et al. 2018; see also https://bigdata.fairness.io.

91. Collins 2002.

92. Eubanks 2017.

93. Crawford 2016.

94. Crawford 2016.

95. Buolamwini 2017.

96. See https://www.ajlunited.org.

97. See http://www.fatml.org.

98. Lorica 2018.

99. Keyes 2018.

100. Hoffman 2019.

101. Collins 2002, 297.

102. Collins 2002, 297.

103. Benjamin 2019a.

104. Angwin et al. 2016.

105. For example, see the most recent FAT* conference program at https://fatconference.org/2019/program.html, although the conversation there is beginning to grapple with tensions between fair decision making and the long-term goals of social equality, as in Mouzannar, Ohannessian, and Srebro 2019.

106. Lewis et al. 2018.

107. Irani et al. 2010.

108. Srinivasan 2017.

109. Escobar 2018.

110. Subcomandante Marcos 2000.

111. See https://datasociety.net, https://ainowinstitute.org, https://www.newschool.edu/digital-equity-lab, https://datajusticelab.org, and https://publicdatalab.org.

112. See https://chupadados.codingrights.org.

113. See https://www.fatml.org, https://datasociety.net, https://civic.mit.edu, https://datajusticelab.org, http://www.communitysolutionsva.org/files/Building_Consentful_Tech_zine.pdf, https://www.odbproject.org, and http://femtechnet.org/about/the-network.

114. See https://alliedmedia.org/amc2018/design-justice-track.

2 Design Practices

1. Wakabayashi 2017.

2. For refutations of the memo's arguments, see Sadedin 2017; Fuentes 2017; Johnson 2017; and Barnett and Rivers 2017. Eagly (2017) argued that there is some support for the memo's claims about biological differences between men and women, but not for the author's conclusions about diversity policies. Some scholars supported the memo; for an attempt to summarize scientific arguments on both sides, see Stevens and Haidt 2017. For an overview, see Molteni and Rogers 2017.

3. Wiener 2017; Bogost 2017.

4. Zaleski 2017.

5. Waxman 2017.

6. Shetterly 2017.

7. Volz 2017.

8. Angwin et al. 2016.

9. Eubanks 2018.

10. Lyons, It's Going Down, and Bromma 2017; see https://www.politicalresearch.org/2017/01/20/ctrl-alt-delete-report-on-the-alternative-right.

11. Tyson and Maniam 2016; CNN 2016.

12. Papanek 1974.

13. Wajcman 1991; and see Chanda Prescod-Weinstein's "Decolonising Science Reading List" at https://medium.com/@chanda/decolonising-science-reading-list-339fb773d51f; see also Beatrice Martini's "Decolonizing Technology: A Reading List" at https://beatricemartini.it/blog/decolonizing-technology-reading-list.

14. National Center for Women & Information Technology 2018.

15. Nafus, Leach, and Krieger 2006, cited in Dunbar-Hester 2014.

16. See the report *Breaking the Mold: Investing in Racial Diversity in Tech,* http://breakingthemold.openmic.org.

17. See the *Mother Jones* exposé "Silicon Valley Firms Are Even Whiter and More Male Than You Thought," by Josh Harkinson (2014), based on data gathered through FOIA requests. Later, Google released its own data at http://googleblog.blogspot.com/2014/05/getting-to-work-on-diversity-at-google.html. See also Swift 2010.

18. See Thurm 2018.

19. Skinner 2006.

20. Kleiman, n.d.

21. Google 2014.

22. Silbey 2018; Hicks 2017.

23. Dunbar-Hester 2017.

24. Weeden, Cha, and Bucca 2016; Wilson 2016; and Arce and Segura 2015.

25. For an excellent review of this literature, see Gardner, n.d.

26. Kushi and McManus 2016.

27. Irani 2015.

28. Herring, 2009.

29. For example, see Kochan et al. 2003.

30. OpenMIC 2017.

31. Hunt, Layton, and Prince 2015.

32. Hunt, Layton, and Prince 2015.

33. Penny 2014.

34. See https://www.usability.gov/what-and-why/user-centered-design.html.

35. For example, see the analysis of the design process for two virtual cities in the Netherlands, by Oudshoorn, Rommes, and Stienstra (2004).

36. Hamraie 2013.

37. See http://contratados.org/.

38. Melendez 2014.

39. Von Hippel 2005.

40. Schmider 2016.

41. It is beyond the scope of this section to more fully explore the arguments about why, under these conditions (unmet user product specifications for specific groups of users), markets often fail to produce new firms that in theory should emerge to cater specifically to unmet user needs. Suffice it to say that as of the time of writing, gender non-conforming people's specific user needs have not been met by dating app markets.

42. Nielsen 2012.

43. See Userforge.com.

44. Guo, Shamdasani, and Randall 2011.

45. Long 2009.

46. Norman 1990, 16.

47. For more on phenomenological variation, see Ihde 1990.

48. Flower et al. 2007.

49. Wittkower 2016, 7.

50. Wittkower 2016, 7.

51. Chris Schweidler from Research Action Design, cofounder of the Research Justice track at AMC, remixed this saying and turned it into a hilarious operating table meme that illustrates it best.

52. Von Hippel 2005; Schuler and Namioka 1993; and Bardzell 2010.

53. See https://airbnb.design/anotherlens.

54. Miller 2017.

55. O'Neil 2016.

56. McCann 2015; and see http://www.buildwith.org.

57. See the web magazine *Model View Culture* at modelviewculture.org for excellent summaries of these critiques.

58. Prashad 2013.

59. Pursell 1993.

60. Schumacher 1999.

61. Turner 2010.

62. Willoughby 1990.

63. Gregory 2003.

64. Asaro 2000.

65. Bannon, Bardzell, and Bødker 2019.

66. Sanoff 2008.

67. Muller 2003.

68. Dunn 2007.

69. Byrne and Alexander 2006.

70. Von Hippel 2005.

71. Eglash 2004.

72. Bar, Weber, and Pisani 2016.

73. Steen 2011.

74. See IDEO's design toolkit at https://www.ideo.com/post/design-kit.

75. Sanders and Stappers 2008.

76. Ries 2011.

77. O'Neil 2013.

78. Asaro 2014, 346.

79. Srinivasan 2017, 117.

80. For example, see the work of Jan Chipchase at Nokia: http://janchipchase.com/content/essays/nokia-open-studios.

81. Bezdek 2013.

82. For a humorously framed sampling of design process diagrams, see https://designfuckingthinking.tumblr.com.

83. Thatcher 1987.

84. Fals-Borda 1987; White 1996.

85. Mathie and Cunningham 2003.

86. Brown 2017.

87. Charlton 1998.

88. Goggin and Newell 2003.

89. Ellcessor 2016.

90. Kafer 2013.

91. From "10 Principles of Disability Justice," by Patty Berne on behalf of Sins Invalid, quoted in Piepzna-Samarasinha 2019, 26–28.

92. Kafer 2013; Piepzna-Samarasinha 2018; and see https://www.sinsinvalid.org.

93. For example, see https://www.d.umn.edu/~lcarlson/atteam/lawsuits.html.

94. Another example of a social movement shifting research and design is the AIDS Coalition to Unleash Power (ACT UP!), which transformed both the state of biomedical research on HIV and the accessibility of treatment through a potent mix of direct action, media savvy, and policy lobbying. See Shepard and Hayduk 2002.

95. The project was funded and advised by Code for America and NetGain.

96. Detailed information about the project methodology is available at https://morethancode.cc; we also analyzed secondary data, such as IRS form 990 data, for over thirteen thousand relevant nonprofits.

97. The report was coauthored by Sasha Costanza-Chock, Maya Wagoner, Berhan Taye, Caroline Rivas, Chris Schweidler, Georgia Bullen, and the T4SJ Project and is available at https://morethancode.cc. See Costanza-Chock et al. 2018.

98. "Charley" (all interviewee names were changed for anonymity), interviewed in Costanza-Chock et al. 2018.

99. "Heiner" and "Hbiki" in Costanza-Chock et al. 2018.

100. "Hardy" in Costanza-Chock et al. 2018.

101. "Lulu" in Costanza-Chock et al. 2018.

102. "Alda" in Costanza-Chock et al. 2018.

103. "Tivoli" in Costanza-Chock et al. 2018.

104. "Gertruda" in Costanza-Chock et al. 2018.

105. "Charley" in Costanza-Chock et al. 2018.

106. "Matija" in Costanza-Chock et al. 2018.

107. Costanza-Chock et al. 2018.

3 Design Narratives

1. ElBaradei 2003.

2. Crawford 2017a.

3. Crawford 2017b; Burckle 2013.

4. Dyer-Witheford 1999.

5. Tarrow 2010; Walgrave and Rucht 2010.

6. Furness 2007.

7. New York Civil Liberties Union 2014.

8. See http://wearemany.com.

9. Jackson, Bailey, and Foucault Welles 2019.

10. In fact, Hirsch worked at the Center for Civic Media, the same research group that I would become affiliated with as an MIT faculty member years later in 2012.

11. Hirsch 2008.

12. Hirsch 2013.

13. Sifry 2012.

14. Hirsch 2013.

15. Hirsch 2013.

16. Dyer, Gregersen, and Christensen 2011.

17. Kelley and Littman 2001.

18. For example, see Carey 1983 and Starr 2004.

19. *Meritocracy* was originally a satirical term, as Robert Frank (2016) argues in *Success and Luck: Good Fortune and the Myth of Meritocracy*.

20. Merton 1968. The myth of meritocracy also provides key ammunition for challenges to affirmative action.

21. Rhode 1991.

22. Rogers 1962.

23. Bar, Weber, and Pisani 2016.

24. Bar, Weber, and Pisani 2016.

25. Von Hippel 2005.

26. Von Hippel 2005, 76.

27. See *The Eureka Myth* by Jessica Silbey (2014), who disentangles the relationships among creativity, innovation, and patent and copyright law.

28. Ferrucci, Shoenberger, and Schauster 2014.

29. Matias 2012.

30. Gupta 2006.

31. Gray 2015.

32. Brock 2018.

33. Jackson, Bailey, and Foucalt Welles 2019, 12.

34. Davenport and Beck 2001.

35. For example, see https://en.wikipedia.org/wiki/Kelvin_Doe.

36. Downing 2000; Rodriguez 2001; and Milan 2013.

37. Gamson and Wolfsfeld 1993.

38. Turner 2010.

39. Terranova 2000.

40. See the excellent literature review on this topic in Santa Ana, López, and Munguía 2010.

41. Cottle 2008.

42. Wood 2014; Della Porta and Reiter 1998.

43. Baudrillard 1995.

44. Kellner 2004.

45. Klein 2003.

46. Kumanyika 2016.

47. González and Torres 2011.

48. Halleck 2002.

49. Costanza-Chock 2011.

50. Costanza-Chock 2012.

51. See midianinja.org.

52. Blevins 2018; Jackson, Bailey, and Foucalt Welles 2019.

53. Taylor 2018.

54. Maxigas in AUTISTICI/INVENTATI 2017, 12.

55. Maxigas in AUTISTICI/INVENTATI 2017, 12.

56. Maxigas in AUTISTICI/INVENTATI 2017, 12.

57. AUTISTICI/INVENTATI 2017.

58. Lopez et al. 2007; Wolfson 2014; and Coleman 2011.

59. Metz 2016.

60. Simon 1996.

61. Schön 1983.

62. Steen 2013, 6.

63. Hoffman, Roessler and Moon 2004.

64. Alexander, cited in Hoffman, Roessler and Moon 2004.

65. Hoffman, Roessler and Moon 2004.

66. Dourish 2010.

67. Smith et al 2016.

68. Benford and Snow 2000, 614.

69. Smith et al. 2016, 23.

70. Hanna-Attisha et al. 2016; Butler, Scammell, and Benson 2016.

71. See 18F's guide at https://lean-product-design.18f.gov/1-discovery-research.

72. See https://lean-product-design.18f.gov/1-discovery-research.

73. Brown 2009.

74. Gates Foundation, n.d.

75. Gates Foundation, n.d.

76. *Economist* online 2012; and see https://www.gatesfoundation.org/Media-Center/Press-Releases/2018/11/Bill-Gates-Launches-Reinvented-Toilet-Expo-Showcasing-New-Pathogen-Killing-Sanitation-Products.

77. Kennedy 2013.

78. Kennedy 2013.

79. Kramer, quoted in Kennedy 2013.

80. See https://www.appropedia.org.

81. Kass 2013.

82. Prasad 2012.

83. Prasad 2012.

84. Prasad 2012.

85. Hurn, Gyi, and Mackareth 2014.

86. Hurn, Gyi, and Mackareth 2014, 7.

87. Hurn, Gyi, and Mackareth 2014, 7.

88. Hurn, Gyi, and Mackareth 2014, 8.

89. Alter 2012.

90. De Decker's exhaustively researched article details the history of human dung removal systems in relationship to agriculture and food systems. In it, De Decker describes the system of human feces and urine removal that operated effectively for about four thousand years in China, where sealed containers were removed from households all over the country and transported to farmlands, at which point they were composted and used as fertilizer. See De Decker 2010.

91. Tong 2017.

92. See https://www.makethebreastpumpnotsuck.com.

93. Hare 2013.

94. See http://www.transhack.org.

95. Downing 2000.

4 Design Sites

1. As described in Tweney 2009.

2. See https://alliedmedia.org/news/2012/03/04/media-go-go-lab-seeking-work-stations-and-skill-sharing-sessions.

3. Nucera et al. 2012.

4. Ito et al. 2009.

5. See https://www.alliedmedia.org/ddjc/discotech.

6. Detroit Digital Justice Coalition 2012a.

7. Allied Media Conference 2012, 112.

8. Detroit Digital Justice Coalition 2012a.

9. Allied Media Conference 2013, 24.

10. See https://codesign.mit.edu/discotechs/countersurveillance-discotechs.

11. Ad Astra Workshop 2014. The flyer for the event invited participants to learn poster design, screenprinting, book binding, and stop-motion animation, among other techniques.

12. Web We Want 2014.

13. See https://codesign.mit.edu/discotechs; for an account of the Oakland Co-op DiscoTech, see Spitzer 2016.

14. Maxigas 2012.

15. Irani 2015.

16. Nelson, Tu, and Hines 2001.

17. Bengry-Howell and Griffin 2007; Calvo 2011.

18. Rose 1994.

19. Henriques 2011.

20. Partridge 2010.

21. Patel 2009.

22. Durham, Cooper, and Morris 2013; see Durham 2014 for more information on hip hop feminism.

23. Gomez-Marquez and Young 2016, 5.

24. Gomez-Marquez 2015.

25. Watkins 2019.

26. Buhr 2016.

27. Shadduck-Hernández et al. 2016.

28. Wallerstein 2011.

29. Federici, 2004.

30. Ross 1997.

31. Smith et al. 2016.

32. Smith et al. 2016, 101; italics added.

33. Maxigas 2012.

34. Grenzfurthner and Schneider n.d.

35. Renzi, personal communication, 2018.

36. Grenzfurthner and Schneider, n.d., 3

37. Grenzfurthner and Schneider, n.d., 4.

38. Turner 2009.

39. Žižek, quoted in Grenzfurthner and Schneider, n.d.

40. AUTISTICI/INVENTATI 2017.

41. Lombana Bermúdez 2018.

42. See https://registro.tecnox.org.

43. Duong, personal communication, 2018; see also Duong 2013 for a discussion of blogging and other DIY cultural practices in the Cuban context.

44. Fernandes 2010.

45. See http://www.midiaetnica.com.br.

46. Mihal 2014.

47. Chan 2014.

48. See http://www.civicinnovationlab.la.

49. Fung and Wright 2001.

50. City of Boston 2015.

51. See http://kendallsquare.org.

52. Cornell Tech 2017.

53. Gordon and Walter 2015.

54. Gordon and Walter 2015.

55. Chun 2005.

56. Escobar 2012.

57. Gordon and Walter 2015, 14.

58. Schudson 1998.

59. See Smith et al. 2016, 102.

60. This is a widespread pattern in global cities, but in some places, wealthy people never left the city centers; in others, they left but aren't interested in coming back.

61. Wikipedia, n.d.

62. Mikhak et al. 2002; Gershenfeld 2008; and Walter-Herrmann and Bueching 2014.

63. See fablabs.io.

64. Gershenfeld, Gershenfeld, and Cutcher-Gershenfeld 2017.

65. Kafer 2013.

66. See http://fab.cba.mit.edu/about/charter.

67. See http://peerproduction.net/wp-content/uploads/2012/07/maxigas-geneology_of_hacklabs_and_hackerspaces_draft.pdf.

68. See https://store.alliedmedia.org/products/how-to-discotech-zine.

69. See https://adainitiative.org/2014/02/18/howto-design-a-code-of-conduct-for-your-community/.

70. See http://aorta.coop/portfolio_page/anti-oppressive-facilitation/.

71. Clay 2013.

72. Smith et al. 2016, 105.

73. Smith et al. 2016; and see Scholz 2013.

74. Terranova 2000.

75. Smith et al. 2016, 105; Scholz 2013; and Soderberg 2013.

76. Smith et al. 2016, 105.

77. Smith et al. 2016, 106.

78. Benkler 2006, 60.

79. See Gershenfeld, Gershenfeld, and Cutcher-Gershenfeld 2017.

80. Holman 2015.

81. Smith et al. 2016, 108.

82. Smith et al. 2016, 118.

83. Smith et al. 2016, 119–120. As they put it: "Vested economic interests, positions of political authority, cultural privileges, social norms, technological infrastructures and research agendas selectively appropriate the innovative ideas and practices emerging from community workshops. At the moment, a kind of crowd-funded, Silicon Valley social entrepreneurship predominates in workshops, and that frames developments accordingly. [Workshops may be] reduced to specific design issues … without attention to the wider causes and consequences of alternative development pathways. … The question is, can the workshop movement move beyond its demonstrated possibilities for prototyping and become involved in processes for catalyzing deep-seated transformation? … Where workshops try to connect with community activism for social change, as with the Ateneus in Barcelona, or with FabLab in Amersfoort, effort is required to make design, prototyping, and fabrication tools meaningful for the grassroots activists and their causes. Relevance needs to be demonstrated, and not assumed" (Smith et al. 2016, 120).

84. Smith et al. 2016, 110–111.

85. See https://web.archive.org/web/20171115182528/http://www.techshop.ws/tech shop.pdf/.

86. Briscoe and Mulligan 2014.

87. Charlie DeTar has written about hurricane hackers, and how hackathons do produce community but don't typically produce new working technologies or tools, let alone "solve problems." See DeTar 2013a.

88. Zukin and Papadantonakis 2017.

89. Zukin and Papadantonakis 2017.

90. DeTar 2013a.

91. Broussard 2018.

92. Lin 2016.

93. Lin 2016.

94. DeTar 2013a.

95. "Becca," "Joss," and "Tal," interviewed for #MoreThanCode in Costanza-Chock et al. 2018. See https://bit.ly/morethancode-keytakeaways and https://morethancode .cc/quotes to explore key findings and pull quotes from practitioners.

96. "Erica" and "Heiner," in Costanza-Chock et al. 2018.

97. "Heiner," in Costanza-Chock et al. 2018.

98. "Elioenai," in Costanza-Chock et al. 2018.

99. "Matthew," in Costanza-Chock et al. 2018.

100. Robinson and Johnson 2016.

101. "Hardy," "Tal," and "Joss," in in Costanza-Chock et al. 2018.

102. "Manuel" and "Margerta," in Costanza-Chock et al. 2018.

103. "Isaac," in Costanza-Chock et al. 2018.

104. "Ivar" and "Luna," in Costanza-Chock et al. 2018.

105. "Luna," in Costanza-Chock et al. 2018.

106. Grenzfurthner and Schneider 2009.

107. Toupin 2014.

108. Henry 2014; see also Fox, Ulgado, and Rosner 2015.

109. See https://www.facebook.com/sugarshackLA.

110. Smith et al. 2016; Hielscher 2015.

111. Smith et al. 2016.

112. Selvaraj 2016.

113. Smith et al. 2016, 122.

114. See Palfrey 2015; Lee and Phillips 2018.

115. Resnick and Rusk 1996; Resnick, Rusk, and Cooke 1998. See also Jaleesa Trapp's work as an educator, activist, creative learning advocate, and former Tacoma Clubhouse Coordinator. See https://www.media.mit.edu/people/jaleesat/updates.

116. Juris 2008.

117. See Costanza-Chock 2003 and Sreberny 2004.

118. Dichter 2004.

119. Costanza-Chock 2012.

120. See https://lesbianswhotech.org.

121. See http://www.transhack.org.

122. See https://whoseknowledge.org.

123. See https://hackathon.inclusivedesign.ca.

124. D'Ignazio et al. 2016.

125. Lin 2016.

126. Lin 2016.

127. Richard et al. 2015.

128. National Center for Women & Information Technology, quoted in Richard et al. 2015, 115.

129. "Tom," in Costanza-Chock et al. 2018.

130. "Odell," in Costanza-Chock et al. 2018.

131. "Landon" and "Odell," in Costanza-Chock et al. 2018.

132. Costanza-Chock et al. 2018.

133. Lorde 1984.

5 Design Pedagogy

1. Harvey 2008.

2. Baptiste 2014.

3. Joint Center for Housing Studies of Harvard University 2018.

4. See http://www.clvu.org/our_history.

5. See https://righttothecity.org.

6. Leyba et al. 2013.

7. Aristotle, cited in Halliwell 1986.

8. Freire 2018.

9. Mayo 1999.

10. Highlander Research and Education Center 1997.

11. For the full statement, see https://progressivetech.org/blog/2018/02/27/movement-tech-statement.

12. See https://projectsouth.org.

13. See https://www.movementhistory.org.

14. Costanza-Chock 2014. See also http://idepsca.org.

15. See https://www.myalia.org.

16. Center for Urban Pedagogy 2011.

17. See https://detroitcommunitytech.org/?q=learning-materials.

18. For an excellent recent summary of a parallel ongoing scholarly conversation about digital media and literacy, see Hobbs 2016.

19. Wagoner 2017, 12.

20. Wagoner 2017, 12.

21. See https://walkerart.org/magazine/never-not-learning-summer-specific-part-1-intro-and-identities.

22. See https://www.cmu.edu/qolt/.

23. Ding, Cooper, and Pearlman 2007.

24. According to the authors, one PAD student developed an interesting method for gathering user requirements and possible solutions from wheelchair users in India: the student provided end users with a camera, and asked them to document mobility barriers that they encountered in daily life and then fill out a form that included open-ended comments, an accessibility scale, and a space for suggested improvements to both the built environment and to the assistive device.

25. D'Ignazio and Klein 2019.

26. See http://eqxdesign.com; Smyth and Dimond 2014; and http://designjustice network.org.

27. See https://databasic.io/en/; see also Bhargava and D'Ignazio 2015.

28. See https://www.alliedmedia.org/ddjc/discotech and https://databasic.io/en/ culture.

29. See http://openstreetmap.org; https://publiclab.org; http://mapafeminicidios .blogspot.mx/p/inicio.html; civic.mit.edu/2013/08/07/the-detroit-geographic-expe dition-and-institute-a-case-study-in-civic-mapping; and https://www.propublica.org/ article/lost-mothers-maternal-health-died-childbirth-pregnancy.

30. See http://rapresearchlab.com; for data murals, see https://datatherapy.org and see also Bhargava et al. 2016.

31. Papert and Harel 1991.

32. Piaget, cited in Sabelli 2008.

33. Boud and Feletti 2013.

34. Wilson 1996.

35. Resnick et al. 2009.

36. Resnick, Rusk, and Cooke 1998.

37. Resnick, Rusk, and Cooke 1998.

38. Bruckman and Resnick 1996.

39. Levitt 2017.

40. See https://www.decolonisingdesign.com.

41. Margolin 1996, 3.

42. Margolin 1996, 5.

43. See hooks 1994, 148.

44. I taught the course five times; while I was on leave during the spring of 2013, it was taught by Federico Casalegno, with graduate student Denise Cheng.

45. Scholz and Schneider 2016.

46. All project case studies are available at https://codesign.mit.edu/projects.

47. Leyba et al. 2013.

48. Crockford et al. 2014.

49. McGregor et al. 2013.

50. See the blank template working agreement that we use in the Codesign Studio, available at http://bit.ly/codesign-agreement-template.

51. Racin and Gordon 2018.

52. Duncan et al. 2013, 22.

53. Duncan et al. 2013, 22. In addition to the importance of written working agreements with a community partner, teams in the Codesign Studio also emphasize the need to create an ecosystem map at the beginning of the project to better understand all of the players in the space, to respect community partners' decisions about what to prioritize for prototyping and validation, and to be wary of the dynamics of appropriation (CCTV 2013).

54. Design Studio for Social Intervention 2013.

55. Henderson et al. 2017.

56. Wu et al. 2017.

57. Fernandez et al. 2014.

58. Leyba et al. 2013.

59. Weishaar, Zhong, and Cheng 2017.

60. Fernandez et al. 2014.

61. For example, Goldschmidt (2003) analyzed design education in architectural schools and found that students primarily desired a focus on form and creativity, while paying less attention to, and in some cases actively resenting, seminars, instructors, and crits (desk critiques) that focus on real-world aspects of architecture. Students expected design education to be an area where they were able to give free reign to their creative impulses and produce architectural models and concepts that were formally interesting and aesthetically appealing, creative, unique, or new. They disparaged those aspects of design education that emphasize how real-world architectural structures or spaces are always linked to a particular location with people, history, culture, environment, and so on. In part, the author traced this to the *star culture* of world-famous architects, which permeates the atmosphere of design schools and student aspirations despite the reality of the actual job market, let alone the larger structural questions of the long-term unsustainability of corporate megastructures.

62. See https://aorta.coop/resources.

63. Chakravartty 2006.

64. Fernandez et al. 2014.

65. Freeman 1972.

66. Mohammad et al. 2016.

67. D'Ignazio and Klein 2019.

68. Jordan et al. 2016.

69. Asharia et al. 2013.

70. Wu et al. 2017.

71. Mohammad et al. 2016.

72. Delazari et al. 2016.

73. Fernandez et al. 2014.

74. Shah et al. 2014.

75. Mohammad et al. 2016.

76. Mawson 2003.

77. Shah et al. 2014.

78. Henderson et al. 2017.

79. McGregor et al. 2013.

80. Irani 2015.

81. Design Studio for Social Intervention 2013, 20.

82. Lu et al. 2014.

83. Broussard 2018.

84. McGregor et al. 2013.

85. Design Studio for Social Intervention 2013.

86. CCTV 2013.

87. Obama 2016.

88. Kastrenakes 2016.

89. See https://advancementproject.org/issues/stpp.

90. Abraham 2011.

91. Flores 2007.

92. Burdge, Hyemingway, and Licona 2014.

93. Cottom 2017.

94. As Cottom argues, tech and design courses at community colleges are crucial to enable low-income people to develop their knowledge and skills.

95. Reich and Ito 2017.

96. See http://www.exploringcs.org.

97. See https://code.org/diversity.

98. See https://tsl.mit.edu/projects/swipe-right.

99. Resnick 2017; although somewhat ironically, Ito and Reich (2017) found that Scratch's approach to learning (open-ended, minimally guided, student-driven) may disproportionally benefit students who are already the most advantaged.

100. To take just one of many examples, a group of philosophers and computer scientists at Harvard recently developed a set of learning modules called Embedded EthiCS, designed to integrate ethical reasoning into a CS curriculum and teach students about how to consider the ethical implications of their work, as well as how to decide what technologies should be built or refused (see http://embeddedethics.seas.harvard.edu).

101. See https://www.ncwit.org, and the Wikipedia category "Organizations for Women in Science and Technology" (http://en.wikipedia.org/wiki/Category:Organizations_for_women_in_science_and_technology); for a recent review of best practices in inclusive computer science education, see Hamilton et al. 2016.

102. See debianwomen.org, geekfeminism.org, pyladies.net, http://www.blackgirlscode.com, and the Wikipedia category "Organizations for Women in Science and Technology" (http://en.wikipedia.org/wiki/Category:Organizations_for_women_in_science_and_technology).

103. See http://www.blackgirlscode.com.

104. See http://girlswhocode.com.

105. See http://www.code2040.org.

106. In *Education and Work*, Du Bois attacked trade schools on their own terms, for continuing to teach trades that he argued were being rapidly displaced by the larger reorganization of work, automation, factories, and the rise of multinational firms. This double critique remains surprisingly relevant. Is it really true that we can expect continued growth in well-paying coding jobs? Many factors militate against this possibility: these include outsourcing, automation, and increased competition for the well-paying coding jobs that do exist. Du Bois describes how, for Booker T. Washington and other advocates of the trade schools, the goal was to train Black people for employment; this was meant to provide the foundation for the creation of Black wealth and ultimately lead to the uplift of all Black people and eventually to integration with white society. Du Bois, on the other hand, wanted higher

education to become an institution that would train Black people for leadership, vision, and moral and cultural excellence. He also wanted college graduates to be prepared to take on key roles at the highest levels of industry and science.

107. Du Bois 1932, 61. See also Du Bois 1903, 63. Du Bois said "men," not people.

Directions for Future Work

1. Luo 2018, 5.

2. Godz 2018.

3. See https://www.icrac.net/open-letter-in-support-of-google-employees-and-tech-workers.

4. Child and family detentions increased under the Obama administration as well, as documented by Detention Watch Network and other immigrant rights organizations, but the Trump administration took these policies to new heights of cruelty. Detention Watch Network, n.d.

5. Smith and Bogado 2018; Human Rights Watch 2018.

6. Chao 2018. MIT faculty members (I was a coauthor) also circulated an open letter from scholars and scientists in support of the Microsoft workers' campaign; the letter was signed by nearly five hundred faculty, scientists, and researchers across the country. It is available at https://actionnetwork.org/petitions/an-open-letter-to-microsoft-drop-your-194-million-ice-tech-contract.

7. Captain 2018.

8. Kauffman 2018.

9. Sydell 2018.

10. Forsythe and Bogdanich 2018.

11. Gallagher 2018; Kottasová 2018.

12. Condliffe 2018.

13. Bright 2018.

14. See https://www.ibmpetition.org.

15. Segarra 2018.

16. See https://techworkerscoalition.org.

17. Science for the People 2018a.

18. Science for the People 2018b.

19. Allen 2019.

20. Gilpin 2015.

21. Moore 2009.

22. Schuler and Namioka 1993.

23. Wolfson 2014.

24. Smith et al. 2016.

25. Braman 2011; Braman 2012.

26. See https://www.thejustdatalab.com/resources.

27. Koopmans 2004.

28. Dyer-Witheford 1999.

29. Browne 2015.

30. Roston 2017; Brian Resnick 2017.

31. For example, Ben Green's (2019) book the *Smart Enough City* will hopefully lead to a wave of actionable critiques of so-called smart city discourse, policy, and practices.

32. Costanza-Chock 2018.

33. Flyvbjerg 2005.

34. DiSalvo 2012, 118.

35. Coffey 2015.

36. Mills 2015; Alexander 2005.

37. Noble 2018, 171–172.

38. See https://criticalracedigitalstudies.com.

39. Khalil and Kier 2017.

40. Brock 2018.

41. Friedman and Nissenbaum 1996.

42. Lee et al. 2016.

43. ADA 2007.

44. Hamraie 2017.

45. Bush 1983.

46. Raji and Buolamwini 2019.

47. See https://bostoncivic.media.

48. Here, Fox Harrell's work on phantasmal media (2013) and Sandra Braman's work on identity and the information state (2009) provide extremely relevant touchstones.

49. Connell et al. 1997.

50. See http://www.iccsafe.org.

51. See https://www.iso.org/standard/52075.html.

52. Varon and Cath 2015; ten Oever 2018.

53. Braman 2012.

54. See, for example, https://en.wikipedia.org/wiki/Accessibility.

55. However, law can be a very slow and sometimes blunt instrument for shaping technology design. For example, a recent paper by Goodman and Flaxman (2016) discusses the push for a *right to an explanation* or *algorithmic transparency* law. The authors argue that transparency as an approach will not address the most important algorithmic harms for several reasons: first, they claim that the right to an explanation in the EU General Data Protection Regulation (GDPR), if conceived of as a requirement meaning that algorithm makers must disclose "how their algorithms function," will be both overbroad and impossible to fulfill. The authors describe how common machine-learning techniques do not produce decision-making processes that are explainable in the common meaning of the term *explanations*, as in "meaningful information about the logic of processing." In other words, computer scientists who use machine learning to create algorithms often can only provide explanations of this kind within a very specific kind of limited query.

56. Federici 2012.

57. Purao, Bagby, and Umapathy 2008.

58. Cizek et al. 2019.

59. Braman 2012.

60. Purao, Bagby, and Umapathy 2008.

61. DeTar 2013b.

62. Bernal 1998; Collins 2002; and Harding 2004.

63. Broussard 2018.

64. Dyer-Witheford 1999.

65. Scholz and Schneider 2016.

66. See https://colloqate.org/design-justice-summit.

67. See http://eqxdesign.com.

68. Racin and Gordon 2018.

69. See https://foundation.mozilla.org/en/initiatives/responsible-cs/challenge.

70. Spelic 2018.

71. Spelic 2018.

References

18F. 2017. *Lean Product Design*. Accessed October 2, 2017. https://lean-product-design.18f.gov/1-discovery-research.

99% Invisible. 2014. "A Short History of the High Heel." *Slate*, June 19, 2014. http://www.slate.com/blogs/the_eye/2014/06/19/_99_invisible_roman_mars_the_gender_bending_history_of_the_high_heel.html.

Abraham, Mark. 2011. "New Reports Highlight Potential Policy Solutions to Connecticut Achievement Gap." *DataHaven*, January 23, 2011. http://ctdatahaven.org/blog/new-reports-highlight-potential-policy-solutions-connecticut-achievement-gap.

Ad Astra Workshop. 2014. "Next Ad Astra Workshop: At the Social Justice Discotech Conference on A24!" Save the date flyer. April 3, 2014. https://adastracomix.com/2014/04/03/next-ad-astra-workshop-at-the-social-justice-discotech-conference-on-a24.

ADA. 2007. "Website Accessibility under Title II of the ADA." *ADA Best Practices Tool Kit for State and Local Governments*. May 7, 2007. https://www.ada.gov/pcatoolkit/chap5toolkit.htm.

Adam, Emma K., Jennifer A. Heissel, Katharine H. Zeiders, Jennifer A. Richeson, Emily C. Ross, Katherine B. Ehrlich, Dorainne J. Levy, et al. 2015. "Developmental Histories of Perceived Racial Discrimination and Diurnal Cortisol Profiles in Adulthood: A 20-Year Prospective Study." *Psychoneuroendocrinology* 62 (December): 279–291.

Adamoli, Ginevra. 2012. "Social Media and Social Movements: A Critical Analysis of Audience's Use of Facebook to Advocate Food Activism Offline." Electronic theses, treatises, and dissertations, paper 5310. Doctoral Dissertation for The Florida State University College of Communication and Information. http://citeseerx.ist.psu.edu/viewdoc/download?doi=10.1.1.660.742&rep=rep1&type=pdf.

Alexander, M. Jacqui. 2005. *Pedagogies of Crossing: Meditations on Feminism, Sexual Politics, Memory, and the Sacred*. Durham, NC: Duke University Press.

Aliseda, Atocha. 2006. *Abductive Reasoning*. Vol. 330. Dordrecht: Springer, 2006.

Allen, Jonathan, ed. 2019. *March 4: Scientists, Students, and Society*. Anniversary edition. Cambridge, MA: MIT Press.

Allied Media Conference. 2012. "DiscoTechs UNITE! Part 2." Workshop description. Allied Media Conference Program. https://www.alliedmedia.org/sites/tmpstage.dev.altissima.theworkdept.com/files/amc_program2012v6_061212.pdf.

Allied Media Conference. 2013. "Discovering Technology (DiscoTech) Lab." Workshop description. Allied Media Conference Program. https://www.alliedmedia.org/files/amc2013_program_0_1.pdf.

Allied Media Projects. n.d. "Network Principles." Accessed June 15, 2019. https://www.alliedmedia.org/about/network-principles.

Alpert, Meryl. 2018. *Giving Voice: Mobile Communication, Disability, and Inequality*. Cambridge, MA: MIT Press.

Alter, Lloyd. 2012. "Crapping on Bill Gates' 'Reinvent the Toilet' Winner." Treehugger, August 22, 2012. https://www.treehugger.com/bathroom-design/crapping-bill-gates-reinvent-toilet-winner.html.

Angwin, Julia, Jeff Larson, Surya Mattu, and Lauren Kirchner. 2016. "Machine Bias." *ProPublica*, May 23, 2016. https://www.propublica.org/article/machine-bias-risk-assessments-in-criminal-sentencing.

Angwin, Julia, and H. Grassegger. 2017. "Facebook's Secret Censorship Rules Protect White Men from Hate Speech but Not Black Children," *ProPublica*, June 27, 2017. https://www.propublica.org/article/facebook-hate-speech-censorship-internal-documents-algorithms.

Arce, Eric, and Denise A. Segura. 2015. "Stratification in the Labor Market." In *The Wiley Blackwell Encyclopedia of Race, Ethnicity, and Nationalism*, edited by John Stone, Rutledge Dennis, Polly Rizova, Anthony Smith, and Xiaoshuo Hou. Chichester, West Sussex, UK: John Wiley & Sons. https://doi.org/10.1002/9781118663202.wberen226.

Arnstein, Sherry R. 1969. "A Ladder of Citizen Participation." *Journal of the American Institute of Planners* 35, no. 4: 216–224.

Asaro, Peter M. 2000. "Transforming Society by Transforming Technology: The Science and Politics of Participatory Design." *Accounting, Management and Information Technologies* 10, no. 4: 257–290.

Asaro, P. 2014. "Participatory Design." In *The Encyclopedia of Science, Technology and Ethics*, 2nd ed., edited by Carl Mitcham, 345–347. New York: Macmillon. http://peterasaro.org/writing/Asaro%20Participatory_Design.pdf.

Asharia, Anjum, Marisa Jahn, Natalicia Tracy, Rafaela Serrano, Leo Burd, Tamer Zoubi, Alexandre Goncalves, et al. 2013. *Claro que Si: Domestic Worker Health & Safety Hotline*. Civic Media Collaborative Design Studio. Cambridge, MA: MIT. https://docs.google.com/document/d/1VU9PPlE4UkhdDGkYZaFehCYGPHVDNHhG4qCM8v77ORE.

Autistici/Inventati. 2017. *+KAOS: Ten Years of Hacking and Media Activism*. Amsterdam: Institute of Network Cultures.

Bailey, Moya Z., Brooke Foucault Welles, and Sarah J. Jackson. 2019. *#HashtagActivism: Race and Gender in America's Network Counterpublics*. Cambridge, MA: MIT Press.

Bannon, Liam, Jeffrey Bardzell, and Susanne Bødker. 2019. "Reimagining Participatory Design." *Interactions* 26, no. 1: 26–32.

Baptiste, Nathalie. 2014. "Staggering Loss of Black Wealth Due to Subprime Scandal Continues Unabated." *American Prospect*, October 13, 2014. http://prospect.org/article/staggering-loss-black-wealth-due-subprime-scandal-continues-unabated.

Bar, François, Matthew S. Weber, and Francis Pisani. 2016. "Mobile Technology Appropriation in a Distant Mirror: Baroquization, Creolization, and Cannibalism." *New Media & Society* 18, no. 4: 617–636. https://doi.org/10.1177/1461444816629474.

Barassi, Veronica. 2013. "Ethnographic Cartographies: Social Movements, Alternative Media and the Spaces of Networks." *Social Movement Studies* 12, no. 1: 48–62.

Bardzell, Shaowen. 2010. "Feminist HCI: Taking Stock and Outlining an Agenda for Design." In *Proceedings of the SIGCHI Conference on Human Factors in Computing Systems*, 1301–1310. New York: ACM.

Barnett, Rosalind, and Caryl Rivers. 2017. "We've Studied Gender and STEM for 25 Years. The Science Doesn't Support the Google Memo." *Recode*, August 11, 2017. https://www.recode.net/2017/8/11/16127992/google-engineer-memo-research-science-women-biology-tech-james-damore.

Baudrillard, Jean. 1995. *The Gulf War Did Not Take Place*. Bloomington: Indiana University Press.

Baughman, Galen, Josh Gravens, Andrew Extein, Susanna Pho, and Miho Kitagawa. 2014. *I Am Not a Dot: Humanizing the US Sex Offender Registry*. Civic Media Collaborative Design Studio. Cambridge, MA: MIT. https://docs.google.com/document/d/1Das6T2VkUad8dLJw-Km56P1H4XA6Q3RTLAjUKW7rIZ0/edit#.

Beauchamp, Toby. 2009. "Artful Concealment and Strategic Visibility: Transgender Bodies and U.S. State Surveillance after 9/11." *Surveillance & Society* 6, no. 4: 356–366.

Benford, Robert D., and David A. Snow. 2000. "Framing Processes and Social Movements: An Overview and Assessment." *Annual Review of Sociology* 26:611–639.

Bengry-Howell, Andrew and Christine Griffin. 2007. "Self-Made Motormen: The Material Construction of Working-class Masculine Identities through Car Modification." *Journal of Youth Studies* 10, no. 4: 439–458.

Benjamin, Ruha. 2016a. "Catching Our Breath: Critical Race STS and the Carceral Imagination." *Engaging Science, Technology, and Society* 2:145–156.

Benjamin, Ruha. 2016b. "Innovating Inequity: If Race Is a Technology, Postracialism Is the Genius Bar." *Ethnic and Racial Studies* 39, no. 13: 2227–2234.

Benjamin, Ruha. 2019a. *Race After Technology: Abolitionist Tools for the New Jim Code.* Cambridge: Polity.

Benjamin, Ruha. 2019b. *Captivating Technology: Race, Carceral Technoscience, and Liberatory Imagination in Everyday Life.* Durham, NC: Duke University Press.

Benkler, Yochai. 2006. *The Wealth of Networks: How Social Production Transforms Markets and Freedom.* New Haven, CT: Yale University Press.

Bernal, Dolores Delgado. 1998. "Using a Chicana Feminist Epistemology in Educational Research." *Harvard Educational Review* 68, no. 4: 555–583.

Bezdek, Barbara L. 2013. "Citizen Engagement in the Shrinking City: Toward Development Justice in an Era of Growing Inequality." *St. Louis University Public Law Review* 33:2014–2020.

Bhargava, Rahul, and Catherine D'Ignazio. 2015. "Designing Tools and Activities for Data Literacy Learners." Paper presented at the Web Science: Data Literacy Workshop. Oxford, UK, April 2015. https://www.media.mit.edu/publications/designing-tools-and-activities-for-data-literacy-learners.

Bhargava, Rahul, Ricardo Kadoukai, Emily Bhargava, Guilherme Castro, and Catherine D'Ignazio. 2016. "Data Murals: Using the Arts to Build Data Literacy." *Journal of Community Informatics* 12, no. 3.

Bivens, R. 2017. "The Gender Binary Will Not Be Deprogrammed: Ten Years of Coding Gender on Facebook." *New Media & Society* 19, no. 6: 880–898.

Blevins, Jeffrey L., 2018. "Social Media and Social Justice Movements after the Diminution of Black-Owned Media in the United States." In *Media Across the African Diaspora: Content, Audiences, and Influence*, edited by Omotayo O. Banjo, 191–203. New York: Taylor & Francis.

Bogost, Ian. 2017. "A Googler's Would-Be Manifesto Reveals Tech's Rotten Core." *Atlantic*, August 6, 2017. https://www.theatlantic.com/technology/archive/2017/08/why-is-tech-so-awful/536052.

Boud, David, and Grahame Feletti. 2013. *The Challenge of Problem-Based Learning.* New York: Routledge.

boyd, danah, Karen Levy, and Alice Marwick. 2014. "The Networked Nature of Algorithmic Discrimination." In *Data and Discrimination: Collected Essays*, edited by Seeta Peña Gangadharan, Virginia Eubanks, and Solon Barocas, 53–58. Washington, DC: Open Technology Institute and the New America Foundation.

Braman, Sandra. 2009. *Change of State: Information, Policy, and Power*. Cambridge, MA: MIT Press.

Braman, Sandra. 2011. "The Framing Years: Policy Fundamentals in the Internet Design Process, 1969–1979." *Information Society* 27, no. 5: 295–310.

Braman, Sandra. 2012. "Privacy by Design: Networked Computing, 1969–1979." *New Media & Society* 14, no. 5: 798–814.

Bright, Peter. 2018. "Amazon Staff to Bezos: Stop Selling Tech to Law Enforcement, Palantir." *Ars Technica*, June 22, 2018. https://arstechnica.com/tech-policy/2018/06/amazon-workers-tell-bezos-to-stop-selling-facial-recognition-to-police/.

Briscoe, Gerard, and Catherine Mulligan. 2014. "Digital Innovation: The Hackathon Phenomenon." *Creativeworks London*, working paper no. 6. http://www.creativeworkslondon.org.uk/wp-content/uploads/2013/11/Digital-Innovation-The-Hackathon-Phenomenon1.pdf.

Brock, André. 2018. "Critical Technocultural Discourse Analysis." *New Media & Society* 20, no. 3: 1012–1030. https://doi.org/10.1177/1461444816677532.

Broussard, Meredith. 2018. *Artificial Unintelligence: How Computers Misunderstand the World*. Cambridge, MA: MIT Press.

brown, adrienne maree. 2017. *Emergent Strategy: Shaping Change, Changing Worlds*. Chico, CA: AK Press.

Brown, Tim. 2009. *Change by Design: How Design Thinking Transforms Organizations and Inspires Innovation*. New York: HarperBusiness.

Browne, Simone. 2015. *Dark Matters: On the Surveillance of Blackness*. Durham, NC: Duke University Press.

Bruckman, Amy, and Mitchel Resnick. 1996. "The MediaMOO Project: Constructionism and Professional Community." *Constructionism in Practice: Designing, Thinking, and Learning in a Digital World*, edited by Yasmin B. Kafai and Mitchel Resnick, 207–222. New York: Routledge.

Buhr, Sarah. 2016. "Weed On-Demand Startup Eaze Inhales $13 Million in Funding to Grow into New Markets." *TechCrunch*, October 24, 2016. https://techcrunch.com/2016/10/24/putthatinyourpipeandsmokeit.

Buolamwini, Joy Adowaa. 2017. "Gender Shades: Intersectional Phenotypic and Demographic Evaluation of Face Datasets and Gender Classifiers." MSc diss., Massachusetts Institute of Technology.

Buolamwini, Joy, and Timnit Gebru. 2018. "Gender Shades: Intersectional Accuracy Disparities in Commercial Gender Classification." *Proceedings of Machine Learning Research* 81:1–15.

Burckle, Frederick. 2013. "Civilian Mortality after the 2003 Invasion of Iraq." *Lancet* 381, no. 9870: 877–879.

Burdge, Hilary, Zami T. Hyemingway, and Adela C. Licona. 2014. *Gender Nonconforming Youth: Discipline Disparities, School Push-Out, and the School-to-Prison Pipeline.* San Francisco and Tucson: Gay-Straight Alliance Network and Crossroads Collaborative at the University of Arizona.

Bush, Corlann Gee. 1983. "Women and the Assessment of Technology: To Think, to Be; to Unthink, to Free." In *Machina ex dea: Feminist Perspectives on Technology*, edited by Joan Rothchild, 151–170. New York: Teacher's College Press.

Butler, Lindsey J., Madeleine K. Scammell, and Eugene B. Benson. 2016. "The Flint, Michigan, Water Crisis: A Case Study in Regulatory Failure and Environmental Injustice." *Environmental Justice* 9, no. 4: 93–97.

Byrne, E., and P. M. Alexander. 2006. "Questions of Ethics: Participatory Information Systems Research in Community Settings." In *Proceedings of the 2006 Annual Research Conference of the South African Institute of Computer Scientists and Information Technologists on IT Research in Developing Countries*, edited by Judith Bishop and Derrick Kourie, 117–126. Republic of South Africa: South African Institute for Computer Scientists and Information Technologists.

Calvo, William. 2011. "Lowriders: Cruising the Color Line." PhD diss., Arizona State University.

Cammaerts, Bart. 2015. "Technologies of Self-Mediation: Affordances and Constraints of Social Media for Protest Movements." In *Civic Engagement and Social Media*, edited by Julie Uldam and Anne Vestergaard, 87–110. London: Palgrave Macmillan.

Caplan, Robyn, J. Donovan, L. Hanson, and Jenna Matthews. 2018. "Algorithmic Accountability: A Primer." Washington, DC: Data & Society. https://datasociety.net/wp-content/uploads/2018/04/Data_Society_Algorithmic_Accountability_Primer_FINAL-4.pdf.

Capps, Kriston. 2017. "Almost 200 Firms Are Interested in Building Trump's Wall." *CityLab*, February 28, 2017. https://www.citylab.com/design/2017/02/almost-200-firms-are-interested-in-building-trumps-border-wall/518049.

Captain, Sean. 2018. "Anti-ICE Protesters Descend on Salesforce Tower in San Francisco." *Fast Company*, July 9, 2018. https://www.fastcompany.com/90193049/anti-ice-protesters-descend-on-salesforce-tower-in-san-francisco.

Carey, James W. 1983. "Technology and Ideology: The Case of the Telegraph." *Prospects* 8:303–325.

CCTV. 2013. *Spring 2013 Codesign Studio Booklet.* Civic Media Collaborative Design Studio. Cambridge, MA: MIT. http://web.mit.edu/schock/www/docs/spring13-codesign-booklet.pdf.

Center for Urban Pedagogy. 2011. *Dialed In: A Cell Phone Literacy Toolkit.* Brooklyn, NY: Center for Urban Pedagogy. http://welcometocup.org/Store?product_id=42.

Chakravartty, Paula. 2006. "Who Speaks for the Governed? World Summit on Information Society, Civil Society and the Limits of 'Multi-stakeholderism.'" *Economic and Political Weekly* 41, no. 3 (January 21–27): 250–257.

Chan, Anita Say. 2014. *Networking Peripheries: Technological Futures and the Myth of Digital Universalism.* Cambridge, MA: MIT Press.

Chao, Sharon. 2018. "MIT Professors Spearhead Petition in Support of Microsoft Employees Protesting Contract with ICE." *Tech*, July 26, 2018, https://thetech.com/2018/07/26/microsoft-ice-contract-petition.

Chardronnet, Ewen. 2015. "GynePunk, the Cyborg Witches of DIY Gynecology." *Makery*, June 30, 2015. http://www.makery.info/en/2015/06/30/gynepunk-les-sorcieres-cyborg-de-la-gynecologie-diy.

Charlton, James I. 1998. *Nothing about Us without Us: Disability Oppression and Empowerment.* Berkeley and Los Angeles: University of California Press.

Chemaly, Soraya. 2016. "The Problem with a Technology Revolution Designed Primarily for Men." *Quartz*, March 16, 2016. https://qz.com/640302/why-is-so-much-of-our-new-technology-designed-primarily-for-men.

Chun, Wendy Hui Kyong. 2005. *Control and Freedom: Power and Paranoia in the Age of Fiber Optics.* Cambridge, MA: MIT Press.

City of Boston. 2015. "Mayor Walsh Releases Report Exploring the Creation of Additional Neighborhood Innovation Districts." Mayor's Press Office, September 28, 2015. https://www.cityofboston.gov/news/Default.aspx?id=20356.

Cizek, Katerina, William Uricchio, Juanita Anderson, Maria Aqui Carter, Detroit Narrative Agency, Thomas Allen Harris, Maori Holmes, Richard Lachman, Louis Massiah, Cara Mertes, Sara Rafsky, Michèle Stephenson, Amelia Winger-Bearskin, and Sarah Wolozin. 2019. *Collective Wisdom: Co-creating with Communities, across Disciplines and with Algorithms.* Cambridge, MA: MIT Open Documentary Lab. https://wip.pubpub.org/collectivewisdom.

Clay, Kelly. 2013. "3D Printing Company MakerBot Acquired in $604 Million Deal." *Forbes*, June 19, 2013. https://www.forbes.com/sites/kellyclay/2013/06/19/3d-printing-company-makerbot-acquired-in-604-million-deal/#614e3b1a1ef8.

CNN. 2016. "Exit Polls 2016." https://www.cnn.com/election/2016/results/exit-polls/national/president.

Code2040. n.d. "Background." *Code2040.* Accessed January 14, 2017. http://www.code2040.org/history.

Coffey, Mary K. 2015. "US American Art in the Americas." In *A Companion to American Art*, edited by John Davis, Jennifer A. Greenhill, and Jason D. LaFountain, 281–298. West Sussex, UK: John Wiley & Sons.

Coleman, E. Gabriella. 2011. "Anonymous: From the Lulz to the Collective Action." *The New Everyday: A Media Commons Project.* http://mediacommons.org/tne/pieces/anonymous-traveling-pure-lulz-land-political-territories.

Collins, Patricia Hill. 2002. *Black Feminist Thought: Knowledge, Consciousness, and the Politics of Empowerment.* New York: Routledge.

Combahee River Collective. 1983. "The Combahee River Collective Statement." In *Home Girls: A Black Feminist Anthology*, edited by Barbara Smith, 264–274. New Brunswick, NJ: Rutgers University Press.

Condliffe, Jamie. 2018. "Amazon Urged Not to Sell Facial Recognition Technology to Police." *New York Times*, June 19, 2018. https://www.nytimes.com/2018/06/19/business/dealbook/amazon-facial-recognition.html.

Connell, Bettye Rose, Mike Jones, Ron Mace, Jim Mueller; Abir Mullick, Elaine Ostroff, Jon Sanford, et al. 1997. *The Principles of Universal Design.* Raleigh, NC: The Center for Universal Design, 1997. https://projects.ncsu.edu/ncsu/design/cud/about_ud/udprinciplestext.htm.

Cook, Cynthia M., John J. Howard, Yevgeniy B. Sirotin, Jerry L. Tipton, and Arun R. Vemury. 2019. "Demographic Effects in Facial Recognition and Their Dependence on Image Acquisition: An Evaluation of Eleven Commercial Systems." *IEEE Transactions on Biometrics, Behavior, and Identity Science* 1, no. 1: 32–41.

Cornell Tech. 2017. "Cornell Tech Campus Opens on Roosevelt Island, Marking Transformational Milestone for Tech in NYC." Cornell Tech, September 12, 2017. https://tech.cornell.edu/news/cornell-tech-campus-opens-on-roosevelt-island-marking-transformational-mile.

Costanza-Chock, Sasha. 2003. "WSIS, the Neoliberal Agenda, and Counterproposals from 'Civil Society.'" *Journal of Communication Inquiry* 3, no. 2: 118–139.

Costanza-Chock, Sasha. 2011. "Digital Popular Communication: Lessons on Information and Communication Technologies for Social Change from the Immigrant Rights Movement." *National Civic Review* 100, no. 3: 29–35.

Costanza-Chock, Sasha. 2012. "Mic Check! Media Cultures and the Occupy Movement." *Social Movement Studies* 11, nos. 3–4: 375–385.

Costanza-Chock, Sasha. 2014. *Out of the Shadows, Into the Streets! Transmedia Organizing and the Immigrant Rights Movement.* Cambridge, MA: MIT Press.

Costanza-Chock, Sasha. 2018. "Design Justice: Towards an Intersectional Feminist Framework for Design Theory and Practice." In *Proceedings of DRS 2018*, volume 1, edited by Cristiano Storni, Keelin Leahy, Muireann McMahon, Peter Lloyd, and Erik Bohemia, 529–540. London: Design Research Society. https://ssrn.com/abstract=3189696.

Costanza-Chock, Sasha, Maya Wagoner, Berhan Taye, Caroline Rivas, Chris Schweidler, Georgia Bullen, and the T4SJ Project. 2018. *#MoreThanCode: Practitioners Reimagine the Landscape of Technology for Justice and Equity*. Research Action Design & Open Technology Institute. morethancode.cc.

Costello, Cary Gabriel. 2016. "Traveling while Trans: The False Promise of Better Treatment." *Trans Advocate*, January 3, 2016. http://transadvocate.com/the-tsa-a-binary-body-system-in-practice_n_15540.htm.

Cottle, Simon. 2008. "Reporting Demonstrations: The Changing Media Politics of Dissent." *Media, Culture & Society* 30, no. 6: 853–872.

Cottom, Tressie McMillan. 2017. *Lower Ed: The Troubling Rise of For-Profit Colleges in the New Economy*. New York: New Press.

Crawford, Kate. 2016. "Can an Algorithm Be Agonistic? Ten Scenes from Life in Calculated Publics." *Science, Technology, & Human Values* 41, no. 1: 77–92.

Crawford, Neta. 2017a. "United States Budgetary Costs of Post-9/11 Wars through FY2018: A Summary of the $5.6 Trillion in Costs for the US Wars in Iraq, Syria, Afghanistan and Pakistan, and Post-9/11 Veterans Care and Homeland Security." *Costs of War*. Brown University Watson Institute for International & Public Affairs. https://watson.brown.edu/costsofwar/papers/2017/USBudgetaryCostsFY2018.

Crawford, Neta. 2017b. "Update on the Human Costs of War for Afghanistan and Pakistan, 2001 to mid-2016." *Costs of War*. Brown University Watson Institute for International & Public Affairs. https://watson.brown.edu/costsofwar/files/cow/imce/papers/2016/War%20in%20Afghanistan%20and%20Pakistan%20UPDATE_FINAL_corrected%20date.pdf.

Crenshaw, Kimberlé. 1991. "Mapping the Margins: Intersectionality, Identity Politics, and Violence against Women of Color." *Stanford Law Review* 43, no. 6 (July): 1241–1299.

Crenshaw, Kimberlé. 1989. "Demarginalizing the Intersection of Race and Sex: A Black Feminist Critique of Antidiscrimination Doctrine, Feminist Theory, and Antiracist Politics." *University of Chicago Legal Forum* 1989, no. 1, article 8: 139–167. http://chicagounbound.uchicago.edu/uclf/vol1989/iss1/8.

Critical Art Ensemble. 2008. "Tactical Media at Dusk?" *Third Text* 22, no. 5: 535–548.

Crockford, Kade, Nathan Freitas, and Jeffrey Warren. 2014. *Spidey: Android-Based Stingray Detector*. Civic Media Collaborative Design Studio. Cambridge, MA: MIT. https://docs.google.com/document/d/12D8WoGjNb5OWloTB5gpyJrlhRWvSg6HutsOp3nG4CcM/edit#heading=h.46bqmkevre6q.

Csikszentmihalyi, Mihaly. 1996. *Creativity: Flow and the Psychology of Discovery and Invention*. New York: Harper Collins.

Cutting, Kieran, and Erkki Hedenborg. 2019. "Can Personas Speak? Biopolitics in Design Processes." In *Designing Interactive Systems Conference 2019 Companion*, 153–157. New York: ACM. https://dl.acm.org/citation.cfm?id=3323911.

D'Ignazio, Catherine, Alexis Hope, Becky Michelson, Robyn Churchill, and Ethan Zuckerman. 2016. "A Feminist HCI Approach to Designing Postpartum Technologies: When I First Saw a Breast Pump I Was Wondering if It Was a Joke." In *Proceedings of the 2016 CHI Conference on Human Factors in Computing Systems*, 2612–2622. New York: ACM. https://dl.acm.org/citation.cfm?id=2858036.2858460.

D'Ignazio, Catherine, and Lauren F. Klein. 2019. *Data Feminism*. Cambridge, MA: MIT Press, 2019.

Dalla Costa, Mariarosa, and Selma James. 1972. *Women and the Subversion of the Community*. Bristol, UK: Falling Wall Press.

Davenport, Thomas H., and John C. Beck. 2001. *The Attention Economy: Understanding the New Currency of Business*. Brighton, MA: Harvard Business Press.

Davies, Carole Boyce. 2007. *Left of Karl Marx: The Political Life of Black Communist Claudia Jones*. Durham, NC: Duke University Press.

Davis, Heath Fogg. 2017. *Beyond Trans: Does Gender Matter?* New York: NYU Press.

De Decker, Kris. 2010. "Recycling Animal and Human Dung Is the Key to Sustainable Farming." *Low Tech Magazine*, September 15, 2010.

De Vogue, Ariana, Mary Kay Mallonee, and Emanuella Grinberg. 2017. "Trump Administration Withdraws Federal Protections for Transgender Students." *CNN*, February 22, 2017. http://www.cnn.com/2017/02/22/politics/doj-withdraws-federal-protections-on-transgender-bathrooms-in-schools/index.html.

Delazari, Tais, Rachel Goor, Val Healy, Emilie Reiser, and Maya Wagoner. 2016. *Vida Verde: Streamlining Workflows*. Civic Media Collaborative Design Studio. Cambridge, MA: MIT. https://docs.google.com/document/d/170kzpGGIx6kedte2fCxqg-iIJTKXwxaRb5d_Rs8AoA0.

Della Porta, Donatella, and Herbert Reiter Reiter, eds. 1998. *Policing Protest: The Control of Mass Demonstrations in Western Democracies*. Vol. 6. Minneapolis: University of Minnesota Press.

Design Studio for Social Intervention. 2013. *Upham's Corner Input Collector*. Civic Media Collaborative Design Studio. Cambridge, MA: MIT. http://web.mit.edu/schock/www/docs/spring13-codesign-booklet.pdf.

DeTar, Charles Frederick. 2013a. "Hackathons Don't Solve Problems." *Civic Media* (blog), May 16, 2013. https://civic.mit.edu/2013/5/16/hackathons-dont-solve-problems.

DeTar, Charles Frederick. 2013b. "InterTwinkles: Online Tools for Non-hierarchical, Consensus-Oriented Decision Making." PhD diss., Massachusetts Institute of Technology. https://tirl.org/writing/intertwinkles.

Detention Watch Network. n.d. "Family Detention." *Detention Watch Network*. Accessed August 25, 2018. https://www.detentionwatchnetwork.org/issues/family-detention.

Detroit Digital Justice Coalition. 2012. "DiscoTechs UNITE! Part 1." In AMC 2012 program, 52. https://www.alliedmedia.org/sites/tmpstage.dev.altissima.theworkdept.com/files/amc_program2012v6_061212.pdf.

Detroit Digital Justice Coalition. 2012. *DiscoTech* zine, no. 4 (July). https://www.alliedmedia.org/files/ddjc_zine_4.pdf.

Dewey, John. 1933. *How We Think: A Restatement of the Relation of Reflective Thinking to the Educative Process*. Vol. 8. Lexington, MA: Heath & Co. Publishers.

Dichter, Aliza. 2004. "From Porto Alegre to Cyberspace." Email, August 5, 2004. http://amsterdam.nettime.org/Lists-Archives/nettime-l-0408/msg00011.html.

Ding, Dan, Rory A. Cooper, and Jon Pearlman. 2007. "Incorporating Participatory Action Design into Research and Education." Paper presented at the International Conference on Engineering Education, Coimbra, Portugal, September 3–7, 2007. http://citeseerx.ist.psu.edu/viewdoc/download?doi=10.1.1.487.7323&rep=rep1&type=pdf.

DiSalvo, Carl. 2012. *Adversarial Design*. Cambridge, MA: MIT Press.

DiSalvo, Carl, and Jonathan Lukens. 2009. "Towards a Critical Technological Fluency: The Confluence of Speculative Design and Community Technology Programs." Paper presented at Digital Arts and Culture, Irvine, CA, December 12–15, 2009.

Dizikes, Peter. 2014. "By Any Media Necessary." *MIT News*, November 25, 2014. https://news.mit.edu/2014/book-social-media-political-movements-1125.

Dourish, Paul. 2010. "HCI and Environmental Sustainability: The Politics of Design and the Design of Politics." In *Proceedings of the 8th ACM Conference on Designing Interactive Systems*, 1–10. New York: ACM. http://citeseerx.ist.psu.edu/viewdoc/download?doi=10.1.1.174.4879&rep=rep1&type=pdf.

Downing, John D. H. 2000. *Radical Media: Rebellious Communication and Social Movements*. Thousand Oaks, CA: Sage.

Downing, John D. H. 2003. "The Independent Media Center Movement and the Anarchist Socialist Tradition." In *Contesting Media Power: Alternative Media in a Networked World*, edited by Nick Couldry and James Curran, 243–257. Oxford: Rowman & Littlefield.

Du Bois, William Edward Burghardt. 1903. *The Talented Tenth*. New York: James Pott and Company.

Du Bois, William Edward Burghardt. 1932. "Education and Work." *Journal of Negro Education* 1, no. 1 (April): 60–74.

Dunbar-Hester, Christina. 2014. "Feminists, Geeks and Geek Feminists: Understanding Gender and Power in Technological Activism." UPenn talk at Media Activism Symposium, December 5, 2014.

Dunbar-Hester, Christina. 2017. "Feminists, Geeks, and Geek Feminists: Understanding Gender and Power in Technological Activism." In *Media Activism in the Digital Age*, edited by Victor Pickard and Guobin Yang, 187–204. Abingdon, UK: Routledge.

Duncan, Lucia, Ashwin Balakrishnan, Carrie Liang, Qian Long, Courtney Supple, Alex DeStefano, C. J. Mayo, et al. 2013. *The Flying Z Streaming Radio Player*. Civic Media Collaborative Design Studio. Cambridge, MA: MIT. https://docs.google.com/document/d/1NlL_Ubk9bzXJoekfrTJbCxdzBlWzJWcRK9CRAez3_nI.

Dunn, Christine E. 2007. "Participatory GIS: A People's GIS?" *Progress in Human Geography* 31, no. 5: 616–637.

Duong, Paloma. 2013. "Bloggers Unplugged: Amateur Citizens, Cultural Discourse, and Public Sphere in Cuba." *Journal of Latin American Cultural Studies* 22, no. 4: 375–397.

Durham, Aisha S. 2014. *Home with Hip Hop Feminism: Performances in Communication and Culture (Intersections in Communications and Culture)*. New York: Peter Lang Publishing Group.

Durham, Aisha, Brittney C. Cooper, and Susana M. Morris. 2013. "The Stage Hip-Hop Feminism Built: A New Directions Essay." *Signs: Journal of Women in Culture and Society* 38, no. 3: 721–737.

Dyer, Jeff, Hal Gregersen, and Clayton M. Christensen. 2011. *The Innovator's DNA: Mastering the Five Skills of Disruptive Innovators*. Brighton, MA: Harvard Business Press.

Dyer-Witheford, Nick. 1999. *Cyber-Marx: Cycles and Circuits of Struggle in High-Technology Capitalism*. Champaign: University of Illinois Press.

Dyer-Witheford, Nick. 2016. "Cybernetics and the Making of a Global Proletariat." *Political Economy of Communication* 4, no. 1. http://www.polecom.org/index.php/polecom/article/view/63.

Eagly, Alice. 2017. "Does Biology Explain Why Men Outnumber Women in Tech?" *Conversation*, August 15, 2017. https://theconversation.com/does-biology-explain-why-men-outnumber-women-in-tech-82479.

Economist online. 2012. "Flushed with Pride." *Economist*, August 15, 2012. https://www.economist.com/babbage/2012/08/15/flushed-with-pride.

Eglash, Ron. 2004. "Appropriating Technology: An Introduction." In *Appropriating Technology: Vernacular Science and Social Power*, edited by Ron Eglash, Jennifer L. Croissant, Giovanna Di Chiro, and Rayvon Fouché, vii–xxi. Minneapolis: University of Minnesota Press.

ElBaradei, Mohamed. 2003. "The Status of Nuclear Inspections in Iraq: An Update." International Atomic Energy Agency, March 7, 2003. https://www.iaea.org/newscenter/statements/status-nuclear-inspections-iraq-update.

Ellcessor, Elizabeth. 2016. *Restricted Access: Media, Disability, and the Politics of Participation*. New York: NYU Press.

EquityXDesign. 2016. "Racism and inequity are products of design. They can be redesigned." *Equity Design* (blog), November 15, 2016. https://medium.com/equity-design/racism-and-inequity-are-products-of-design-they-can-be-redesigned-12188363cc6a.

Escobar, Arturo. 2012. "Notes on the Ontology of Design." Sawyer Seminar, *Indigenous Cosmopolitics: Dialogues about the Reconstitution of Worlds*, organized by Marisol de La Cadena and Mario Blaser, October 2012. http://sawyerseminar.ucdavis.edu/files/2012/12/ESCOBAR_Notes-on-the-Ontology-of-Design-Parts-I-II-_-III.pdf.

Escobar, Arturo. 2018. *Designs for the Pluriverse: Radical Interdependence, Autonomy, and the Making of Worlds*. Durham, NC: Duke University Press.

Eubanks, Virginia. 2018. *Automating Inequality: How High-Tech Tools Profile, Police, and Punish the Poor*. New York: St. Martin's Press.

Fals-Borda, Orlando. 1987. "The Application of Participatory Action Research in Latin America." *International Sociology* 2, no. 4 (December): 329–347.

Federici, Silvia. 2004. *Caliban and the Witch*. New York: Autonomedia.

Federici, Silvia. 2012. *Revolution at Point Zero: Housework, Reproduction, and Feminist Struggle*. New York: PM Press.

Fernandes, Sujatha. 2010. *Who Can Stop the Drums? Urban Social Movements in Chávez's Venezuela*. Durham, NC: Duke University Press, 2010.

Fernandez, Maria, Yorman Nunez, Nushelle Silva, Daniel Wang, and Elizabeth Cho. 2014. *Urban Youth Collaborative SMS Survey Initiative*. Civic Media Collaborative Design Studio. Cambridge, MA: MIT. https://docs.google.com/document/d/1SVHCPF9fnHiLVLLl6SY-ikNK-yIPTqJrEsD7UpI6L2w.

Ferrucci, Patrick, Heather Shoenberger, and Erin Schauster. 2014. "It's a Mad, Mad, Mad, Ad World: A Feminist Critique of Mad Men." *Women's Studies International Forum* 47, pt. A (November–December): 93–101.

Flanagan, Mary, Daniel C. Howe, and Helen Nissenbaum. 2008. "Embodying Values in Technology: Theory and Practice." In *Information Technology and Moral Philosophy*, edited by J. Van den Hoven and J. Weckert, 322–353. Cambridge: Cambridge University Press.

Flores, Alfinio. 2007. "Examining Disparities in Mathematics Education: Achievement Gap or Opportunity Gap?" *High School Journal* 91, no. 1: 29–42.

Floridi, Luciano. 2014. *The Fourth Revolution*. Oxford: Oxford University Press.

Flower, Ashley, Matthew K. Burns, and Nicole A. Bottsford-Miller. 2007. "Meta-analysis of Disability Simulation Research." *Remedial and Special Education*, 28, no. 2: 72–79.

Flyvbjerg, Bent. 2005. "Design by Deception: The Politics of Megaproject Approval." *Harvard Design Magazine* 22 (Spring/Summer): 50–59.

Forsythe, Michael, and Walt Bogdanich. 2018. "McKinsey Ends Work with ICE amid Furor over Immigration Policy." *New York Times*, July 9, 2018. https://www.nytimes.com/2018/07/09/business/mckinsey-ends-ice-contract.html.

Fox, Sarah, Rachel Rose Ulgado, and Daniela Rosner. 2015. "Hacking Culture, Not Devices: Access and Recognition in Feminist Hackerspaces." In *Proceedings of the 18th ACM Conference on Computer Supported Cooperative Work & Social Computing*, 56–68. New York: ACM. https://dl.acm.org/citation.cfm?id=2675223.

Frank, Robert. 2016. *Success and Luck: Good Fortune and the Myth of Meritocracy*. Princeton: Princeton University Press.

Freeman, Jo. 1972. "The Tyranny of Structurelessness." *Second Wave* 2, no. 1.

Freire, Paulo. 1972. *Pedagogy of the Oppressed*. Translated by Myra Bergman Ramos. Freiburg, Germany: Herder and Herder.

Friedman, Batya, ed. 1997. *Human Values and the Design of Computer Technology*. Cambridge: Cambridge University Press.

Friedman, Batya, Peter Kahn, and Alan Borning. 2002. *Value Sensitive Design: Theory and Methods*. UW CSE Technical Report 02-12-01. Seattle: University of Washington.

Friedman, Batya, Peter Kahn, Alan Borning, and Alina Huldtgren. 2013. "Value Sensitive Design and Information Systems." In *Early Engagement and New Technologies: Opening Up the Laboratory*, edited by Neelke Doorn, Daan Schuurbiers, Ibo van de Poel, and Michael E. Gorman, 55–95. New York: Springer.

Friedman, Batya, and Helen Nissenbaum. 1996. "Bias in Computer Systems." *ACM Transactions on Information Systems* 14, no. 3: 330–347.

Friedman, Batya, and Helen Nissenbaum. 1997. "Bias in Computer Systems." In *Human Values and the Design of Computer Technology*, edited by Batya Friedman, 21–40. Cambridge: Cambridge University Press.

Friedman, Batya, and David G Hendry. 2019. *Value Sensitive Design: Shaping Technology with Moral Imagination*. Cambridge/London: MIT Press.

Fry, Tony. 2010. *Design as Politics*. Oxford and New York: Berg.

Fuentes, Agustin. 2017. "The "Google Manifesto": Bad Biology, Ignorance of Evolutionary Processes, and Privilege." *PLOS SciComm Blog* (blog), August 14, 2017. http://blogs.plos.org/scicomm/2017/08/14/the-google-manifesto-bad-biology-ignorance-of-evolutionary-processes-and-privilege.

Fung, Archon, and Erik Olin Wright. 2001. "Deepening Democracy: Innovations in Empowered Participatory Governance." *Politics & Society* 29, no. 1: 5–41.

Furness, Zack. 2007. "Critical Mass, Urban Space and Vélomobility." *Mobilities* 2, no. 2: 299–319. https://doi.org/10.1080/17450100701381607.

Gallagher, Ryan. 2018. "Google Struggles to Contain Employee Uproar Over China Censorship Plans." *Intercept*, August 3, 2018. https://theintercept.com/2018/08/03/google-search-engine-china-censorship-backlash.

Gamson, William A., and Gadi Wolfsfeld. 1993. "Movements and Media as Interacting Systems." *Annals of the American Academy of Political and Social Science* 528, no. 1: 114–125.

Gardner, Sue. n.d. "Why Women Leave Tech: What the Research Says." Accessed September 24, 2018. Google doc available at bit.ly/whywomenleavetech ().

Gates Foundation. n.d. "Reinvent the Toilet: Strategy Overview." https://www.gatesfoundation.org/What-We-Do/Global-Growth-and-Opportunity/Water-Sanitation-and-Hygiene/Reinvent-the-Toilet-Challenge.

Gaver, William W. 1991. "Technology Affordances." In *Proceedings of the SIGCHI Conference on Human Factors in Computing Systems*, 79–84. New York: ACM. https://dl.acm.org/citation.cfm?id=108856.

Gehl, Robert W. 2015. "The Case for Alternative Social Media." *Social Media + Society* 1 (2). https://doi.org/10.1177/2056305115604338.

Gerbaudo, Paolo. 2012. *Tweets and the Streets: Social Media and Contemporary Activism*. London: Pluto Press.

Gershenfeld, Neil. 2008. *Fab: The Coming Revolution on Your Desktop—from Personal Computers to Personal Fabrication*. New York: Basic Books.

Gershenfeld, Neil, Alan Gershenfeld, and Joel Cutcher-Gershenfeld. 2017. *Designing Reality: How to Survive and Thrive in the Third Digital Revolution*. New York: Basic Books.

Gibson, James J. 1979. *The Ecological Approach to Visual Perception*. Boston: Houghton Mifflin.

Gillespie, Tarleton. 2018. *Custodians of the Internet: Platforms, Content Moderation, and the Hidden Decisions That Shape Social Media*. New Haven: Yale University Press.

Gilpin, Robert. 2015. *American Scientists and Nuclear Weapons Policy*. Vol. 2064. Princeton, NJ: Princeton University Press.

Godz, Polina. 2018. "Tech Workers versus the Pentagon." *Jacobin Magazine*, June 6, 2018. https://jacobinmag.com/2018/06/google-project-maven-military-tech-workers.

Goggin, Gerard, and Christopher Newell. 2003. *Digital Disability: The Social Construction of Disability in New Media*. Lanham, MD: Roman & Littlefield.

Goldschmidt, Gabriela. 2003. "Expert Knowledge or Creative Spark? Predicaments in Design Education." Paper presented at Design Thinking Research Symposium 6, Expertise in Design, Sydney, Australia, November 17–19, 2003.

Gomez-Marquez, Jose. 2015. "Our New Project: MakerNurse." MIT Little Devices Lab, October 13, 2015. https://littledevices.org/2013/10/15/our-new-project-makernurse.

Gomez-Marquez, Jose, and Anna Young. 2016. "A History of Nurse Making and Stealth Innovation." MIT Little Devices Lab, May 11, 2016.

González, Juan, and Joseph Torres. 2011. *News for All the People: The Epic Story of Race and the American Media*. London: Verso Books.

Goodman, Bryce, and Seth Flaxman. 2016. "European Union Regulations on Algorithmic Decision-Making and a 'Right to Explanation.'" *AI Magazine* 38, no. 3: 2017. arXiv preprint at https://arxiv.org/abs/1606.08813.

Google. 2014. "Women Who Choose Computer Science—What Really Matters: The Critical Role of Encouragement and Exposure." Google for Education, May 26, 2014. https://edu.google.com/pdfs/women-who-choose-what-really.pdf.

Gordon, Eric, and Stephen Walter. 2015. "The Good User/Citizen: Innovation, Technology, and Governmentality." *The Companion to American Urbanism*, ed Joseph Healthcott. New York: Routledge.

Gray, Kishonna L. 2012. "Intersecting Oppressions and Online Communities: Examining the Experiences of Women of Color in Xbox Live." *Information, Communication & Society* 15, no. 3: 411–428.

Gray, Kishonna. 2015. "#CiteHerWork: Marginalizing Women in Academic and Journalistic Writing." Kishonnagray.com, December 28, 2015. http://www.kishonnagray.com/manifestmy-reality/citeherwork-marginalizing-women-in-academic-and-journalistic-writing.

Gray, Mary L., and Siddharth Suri. 2019. *Ghost Work: How to Stop Silicon Valley from Building a New Global Underclass*. New York: Houghton Mifflin Harcourt.

Green, Ben. 2019. *The Smart Enough City: Putting Technology in Its Place to Reclaim Our Urban Future*. Cambridge, MA: MIT Press.

Gregory, Judith. 2003. "Scandinavian Approaches to Participatory Design." *International Journal of Engineering Education* 19, no. 1: 62–74.

Grenzfurthner, Johannes, and Frank Apunkt Schneider. n.d. "Hacking the Spaces." *Monochrom*. http://www.monochrom.at/hacking-the-spaces.

Guo, Frank Y., Sanjay Shamdasani, and Bruce Randall. 2011. "Creating Effective Personas for Product Design: Insights from a Case Study." In *Proceedings of the 4th International Conference on Internationalization, Design and Global Development*, 37–46. Heidelberg and Berlin: Springer-Verlag. https://dl.acm.org/citation.cfm?id=2028774.

Gupta, Anil K. 2006. "From Sink to Source: The Honey Bee Network Documents Indigenous Knowledge and Innovations in India." *Innovations: Technology, Governance, Globalization* 1, no. 3: 49–66.

Haimson, Oliver L., and Anna Lauren Hoffmann. 2016. "Constructing and Enforcing 'Authentic' Identity Online: Facebook, Real Names, and Non-normative Identities." *First Monday* 21, no. 6. https://firstmonday.org/article/view/6791/5521.

Halberstam, Jack. 2018. *Trans*: A Quick and Quirky Account of Gender Variability*. Oakland: University of California Press.

Halleck, DeeDee. 2002. *Hand-Held Visions: The Impossible Possibilities of Community Media*. New York: Fordham University Press.

Halleck, DeeDee. 2003. "Indymedia: Building an International Activist Internet Network." *Media Development* 50, no. 4: 11–14.

Halliwell, Stephen. 1986. *Aristotle's Poetics*. Chicago: University of Chicago Press.

Hamilton, Margaret, Andrew Luxton-Reilly, Naomi Augar, Vanea Chiprianov, Eveling Castro Gutierrez, Elizabeth Vidal Duarte, Helen H. Hu, et al. 2016. "Gender Equity in Computing: International Faculty Perceptions and Current Practices." In *Proceedings of the 2016 ITiCSE Working Group Reports*, 81–102. New York: ACM. https://dl.acm.org/citation.cfm?id=3024911&dl=ACM&coll=DL.

Hamraie, Aimi. 2013. "Designing Collective Access: A Feminist Disability Theory of Universal Design." *Disability Studies Quarterly* 33:1041–5718.

Hamraie, Aimi. 2017. *Building Access: Universal Design and the Politics of Disability*. Minneapolis: University of Minnesota Press.

Hankerson, David, et al. 2016. "Does Technology Have Race?" In *Proceedings of the 2016 CHI Conference Extended Abstracts on Human Factors in Computing Systems*. New York: ACM. https://doi.org/10.1145/2851581.2892578.

Hanna-Attisha, Mona, Jenny LaChance, Richard Casey Sadler, and Allison Champney Schnepp. 2016. "Elevated Blood Lead Levels in Children Associated with the Flint Drinking Water Crisis: A Spatial Analysis of Risk and Public Health Response." *American Journal of Public Health* 106, no. 2: 283–290.

Harding, Sandra G., ed. 2004. *The Feminist Standpoint Theory Reader: Intellectual and Political Controversies*. New York: Routledge.

Hare, Kristen. 2013. "Migrahack Brings Together Journalists, Programmers and Community." Poynter, December 24, 2013. https://www.poynter.org/news/migrahack-brings-together-journalists-programmers-and-community.

Harkinson, Josh. 2014. "Silicon Valley Firms Are Even Whiter and More Male Than You Thought." *Mother Jones* (May 29, 2014). https://www.motherjones.com/media/2014/05/google-diversity-labor-gender-race-gap-workers-silicon-valley.

Harrell, D. Fox. 2013. *Phantasmal Media: An Approach to Imagination, Computation, and Expression*. Cambridge, MA: MIT Press.

Harvey, David. 2008. "The Right to the City." *City Reader* 6:23–40.

Harwell, Drew, and Nick Miroff. 2018. "ICE Just Abandoned Its Dream of 'Extreme Vetting' Software That Could Predict Whether a Foreign Visitor Would Become a Terrorist," *Washington Post*, May 17, 2018. https://www.washingtonpost. com/news/the-switch/wp/2018/05/17/ice-just-abandoned-its-dream-ofextreme-vetting-software-that-could-predict-whether-a-foreign-visitorwould-become-a-terrorist.

Henderson, Ingrid, Kwabena Ofori-Atta, Lucky Bommireddy, Natali Espitia, and Tabia Smith. 2017. *Peas in a Podcast: East Boston Voices*. Civic Media Collaborative Design Studio. Cambridge, MA: MIT. https://docs.google.com/document/d/1774u5aoX8ENETaaiLU0YbLFRPAQNh1yDxduEyPDynKM.

Henriques, Julian. 2011. *Sonic Bodies: Reggae Sound Systems, Performance Techniques, and Ways of Knowing*. New York: Bloomsbury Publishing USA.

Henry, Liz. 2014. "The Rise of Feminist Hackerspaces and How to Make Your Own." *Model View Culture*, February 3, 2014. https://modelviewculture.com/pieces/the-rise-of-feminist-hackerspaces-and-how-to-make-your-own.

Hernández-Ramírez, Rodrigo. 2018. "On Design Thinking, Bullshit, and Innovation." *Journal of Science and Technology of the Arts* 10, no. 3: 2–45.

Herring, Cedric. 2009. "Does Diversity Pay? Race, Gender, and the Business Case for Diversity." *American Sociological Review* 74, no. 2: 208–224.

Hicks, Marie. 2017. *Programmed Inequality: How Britain Discarded Women Technologists and Lost Its Edge in Computing*. Cambridge, MA: MIT Press.

Hielscher, Sabine. 2015. *Fab Lab Amersfoort, De War: An Innovation History*. January 2015. Centre on Innovation and Energy Demand.

Highlander Research and Education Center. 1997. *A Very Popular Economic Education Sampler*. Highlander Research and Education Center.

Hirsch, Tad. 2008. *Contestational Design: Innovation for Political Design*. PhD diss., Massachusetts Institute of Technology.

Hirsch, Tad. 2013. "TXTmob and Twitter: A Reply to Nick Bilton." Public Practice Studio, October 16, 2013. Archived at http://archive.is/5isUZ.

Hobbs R., ed. 2016. *Exploring the Roots of Digital and Media Literacy through Personal Narrative*. Philadelphia: Temple University Press.

Hoffmann, Anna Lauren. 2019. "Where Fairness Fails: Data, Algorithms, and the Limits of Antidiscrimination Discourse." *Information, Communication & Society* 22, no. 7: 900–915. https://doi.org/10.1080/1369118X.2019.1573912.

Hoffman, Robert R., Axel Roesler, and Brian M. Moon. 2004. "What Is Design in the Context of Human-Centered Computing?" *IEEE Intelligent Systems* 19, no. 4: 89–95.

Holman, Will. 2015. "Makerspace: Towards a New Civic Infrastructure." *Places Journal*, November 2015.

Holmes, Kat. 2018. *Mismatch: How Inclusion Shapes Design*. Cambridge, MA: MIT Press.

hooks, bell. 1994. *Teaching to Transgress: Education as the Practice of Freedom*. New York: Routledge.

Huff, Charles, and Joel Cooper. 1987. "Sex Bias in Educational Software: The Effects of Designers' Stereotypes on the Software They Design." *Journal of Applied Social Psychology* 17, no. 6: 519–532.

Human Rights Watch. 2018. "Code Red: Fatal Consequences of Dangerously Substandard Medical Care in Immigration Detention." Human Rights Watch, June 20, 2018. https://www.hrw.org/report/2018/06/20/code-red/fatal-consequences-dangerously-substandard-medical-care-immigration.

Hunt, Vivian, Dennis Layton, and Sara Prince. 2015. "Why Diversity Matters." McKinsey & Company, January 2015. https://www.mckinsey.com/business-functions/organization/our-insights/why-diversity-matters.

Hurley, Amanda Kolson. 2018. "Should Designers Try to Reform Immigrant Detention?" *CityLab*, July 30, 2018. https://www.citylab.com/design/2018/07/should-designers-try-to-reform-immigrant-detention/565439.

Hurn, Karl, Diane Gyi, and Paul Mackareth. 2014. "Reinventing the Toilet: Academic Research Meets Design Practice in the Pursuit of an Effective Sanitation Solution for All." Paper presented at the Industrial Design Society of America Annual International Conference, Austin, TX, August 2014.

Ihde, Don. 1990. *Technology and the Lifeworld: From Garden to Earth*. Bloomington: Indiana University Press.

Inclusive Design Research Center. n.d. "What Do We Mean by Inclusive Design?" Inclusive Design Research Center. Accessed January 14, 2018. https://idrc.ocadu.ca/resources/idrc-online/49-articles-and-papers/443-whatisinclusivedesign.

Irani, Lilly. 2015. "Hackathons and the Making of Entrepreneurial Citizenship." *Science, Technology, & Human Values* 40, no. 5: 799–824.

Irani, Lilly. 2016. "The Hidden Faces of Automation." *XRDS: Crossroads, the ACM Magazine for Students* 23, no. 2: 34–37.

Irani, Lilly. 2018. "Design Thinking: Defending Silicon Valley at the Apex of Global Labor Hierarchies." *Catalyst: Feminism, Theory, Technoscience* 4, no. 1. https://catalystjournal.org/index.php/catalyst/article/view/29638.

Irani, Lilly. 2019. *Chasing Innovation: Making Entrepreneurial Citizens in Modern India*. Princeton, NJ: Princeton University Press, 2019.

Irani, Lilly, Janet Vertesi, Paul Dourish, Kavita Philip, and Rebecca E. Grinter. 2010. "Postcolonial Computing: A Lens on Design and Development." In *Proceedings of the SIGCHI Conference on Human Factors in Computing Systems*, 1311–1320. New York: ACM. https://dl.acm.org/citation.cfm?id=1753522.

Ito, Joichi. 2017. "Resisting Reduction: A Manifesto." *Journal of Design and Science*, last updated December 2, 2018. https://jods.mitpress.mit.edu/pub/resisting-reduction.

Ito, Mizuko, Sonja Baumer, Matteo Bittanti, danah boyd, Rachel Cody, Becky Herr Stephenson, Heather A. Horst, et al. 2009. *Hanging Out, Messing Around, and Geeking Out: Kids Living and Learning with New Media*. Cambridge, MA: MIT Press.

Ito, Mizuko, and Justin Reich. 2017. *From Good Intentions to Real Outcomes: Three Myths about Education Technology and Equity*. Irvine, CA: Digital Media and Learning Research Hub.

Jackson, Sarah, Moya Bailey, and Brooke Foucault Welles. 2019. *#Hashtag Activism: Race and Gender in America's Networked Counterpublics*. Cambridge, MA: MIT Press.

Jencks, Christopher. "Racial Bias in Testing." 1998. In *The Black-White Test Score Gap*, edited by C. Jencks and M. Phillips, 55–85. Washington, DC: Brookings Institution Press.

Jenkins, Henry, Sangita Shresthova, Liana Gamber-Thompson, Neta Kligler-Vilenchik, and Arely Zimmerman. 2016. *By Any Media Necessary: The New Youth Activism*. New York: NYU Press.

Jobin-Leeds, Greg, and AgitArte, eds. 2016. *When We Fight We Win!* New York: New Press. http://www.bruno-latour.fr/sites/default/files/50-MISSING-MASSES-GB.pdf.

Johnson, Stefanie. 2017. "What the Science Actually Says About Gender Gaps in the Workplace." *Harvard Business Review*. https://hbr.org/2017/08/what-the-science-actually-says-about-gender-gaps-in-the-workplace.

Khalil, Deena, and Meredith Kier. 2017. "Critical Race Design: An Emerging Methodological Approach to Anti-racist Design and Implementation Research." *International Journal of Adult Vocational Education and Technology (IJAVET)* 8, no. 2 (2017): 54–71.

Kumanyika, Chenjerai. 2016. "Livestreaming in the Black Lives Matter Network." In *DIY Utopia: Cultural Imagination and the Remaking of the Possible*, edited by Amber Day, 169–188. Lanham, MD: Lexington Books.

Lee, Julia. 2016. "Alarming TSA." *The Fourth Wave*, February 3, 2016. https://thefourthwavepitt.com/2016/02/03/alarming-tsa.

Lee, Una, Nontsikelelo Mutiti, Carlos Garcia, and Wes Taylor, eds. 2016. Principles for Design Justice. *Design Justice Zine*, no. 1. Detroit: Design Justice Network. http://designjusticenetwork.org/zine.

Lee, Victor R., and Abigail Phillips, eds. 2018. *Reconceptualizing Libraries: Perspectives from the Information and Learning Sciences*. New York: Routledge.

Levitt, Janna. 2017. "Dori Tunstall Wants to Decolonize Design Education." *Azure Magazine*, November 1, 2017. https://www.azuremagazine.com/article/elizabeth-dori-tunstall-ocad.

Lewin, Kurt. 1946. "Action Research and Minority Problems." *Journal of Social Issues* 2, no. 4: 34–46.

Lewis, Jason Edward, Noelani Arista, Archer Pechawis, and Suzanne Kite. 2018. "Making Kin with the Machines." *Journal of Design and Science*, July 16, 2018. https://doi.org/10.21428/bfafd97b.

Leyba, Mike, Terry Marshall, Nene Igietseme, Dara Yaskil, CL/VU Boston Met Region Bank Tenant Association Leadership Team, Ed Cabrera, and Triana Kazaleh Sirdenis. 2013. *City Life/Vida Urbana's Carnival Toolkit Co-Design Process Case Study*. Civic Media Collaborative Design Studio. Cambridge, MA: MIT. https://docs.google.com/document/d/1Y0an8-1FLSutN0WJUJbyKq6A6AMil1NWqvdfGm9bzqk.

Lin, Gloria. 2016. "Masculinity and Machinery: Analysis of Care Practices, Social Climate and Marginalization at Hackathons." *Model View Culture*, December 15, 2016. https://modelviewculture.com/pieces/masculinity-and-machinery-analysis-of-care-practices-social-climate-and-marginalization-at-hackathons.

Locke, John. 2014. *Second Treatise of Government: An Essay Concerning the True Original, Extent and End of Civil Government.* West Sussex, UK: John Wiley & Sons.

Lombana Bermúdez, Andrés. 2018. "Economía digital, cultura maker, y nuevas formas de creación de valor." In *Jóvenes, transformación digital y formas de inclusión en América Latina*, edited by Cristóbal Cobo et al., 357–361. Montevideo, Uruguay: Penguin Random House.

Long, Frank. 2009. "Real or Imaginary: The Effectiveness of Using Personas in Product Design." In *Proceedings of the Irish Ergonomics Society Annual Conference.* May 2009, 1–10. Dublin: Irish Ergonomics Society. https://s3.amazonaws.com/media.loft.io/attachments/Long%20(2009)%20Real%20or%20Imaginary.pdf.

Lopez, Alfredo, Jamie McClelland, Eric Goldhagen, Daniel Gillmor, and Amanda B. Hickman. 2007. *The Organic Internet: Organizing History's Largest Social Movement.* Entremundos Publications. https://web.archive.org/web/20111017124432/https://mayfirst.org/sites/default/files/organicinternet.1.5.pdf.

Lorde, Audre. 1984. "The Master's Tools Will Never Dismantle the Master's House." In *Sister Outsider*, 110–114. Berkeley, CA: Crossing Press.

Lorica, Ben. 2018. "We Need to Build Machine Learning Tools to Augment Machine Learning Engineers." O'Reilly, January 11, 2018. https://www.oreilly.com/ideas/we-need-to-build-machine-learning-tools-to-augment-machine-learning-engineers.

Lu, Tiffany, Paulina Haduong, Eva Galperin, Jillian York, Hugh D'Andrade, Mark Burdett, and Alvaro Gil. 2014. *Surveillance Self Defense.* Civic Media Collaborative Design Studio. Cambridge, MA: MIT. https://docs.google.com/document/d/1FeWZ7lqWUu1qj5781Cjj3RmLWZZMg7D2d82jt0f27-M.

Luo, Lauren. 2018. "#TechWontBuildIt: A Comprehensive Investigation of Key Movement Activity and Organizing Tactics." *Networked Social Movements: Media & Mobilization* (CMS.361 course blog), December 18, 2018. Cambridge, MA: MIT. https://networkmovements.wordpress.com/2018/12/18/final-project-post-techwontbuildit.

Lyons, Matthew, Its Going Down, and Bromma. 2017. *Ctrl-Alt-Delete: An Antifascist Report on the Alternative Right.* Montreal: Kersplebedeb Publishing.

Manders-Huits, Noemi. 2011. "What Values in Design? The Challenge of Incorporating Moral Values into Design." *Science and Engineering Ethics* 17, no. 2 (June): 271–287.

Margolin, Victor. 1996. "Teaching Design History." *Statements* 11, no. 2: 5–14. http://victor.people.uic.edu/articles/teachdesignhistory.pdf.

Margolin, Victor. 2006. *The Idea of Design*. Cambridge, MA: MIT Press.

Mathie, Alison, and Gord Cunningham. 2003. "From Clients to Citizens: Asset-Based Community Development as a Strategy for Community-Driven Development." *Development in Practice* 13, no. 5: 474–486.

Matias, J. Nathan. 2012. "Designing Acknowledgment on the Web." MIT Center for Civic Media, November 1, 2012. https://civic.mit.edu/blog/natematias/designing-acknowledgment-on-the-web.

Mawson, Brent. 2003. "Beyond 'The Design Process': An Alternative Pedagogy for Technology Education." *International Journal of Technology & Design Education* 13, no. 2 (May): 117–128.

Maxigas, Peter. 2012. "Hacklabs and Hackerspaces: Tracing Two Genealogies." *Journal of Peer Production*, no. 2. http://peerproduction.net/issues/issue-2/peer-reviewed-papers/hacklabs-and-hackerspaces.

Mayo, Peter. 1999. *Gramsci, Freire and Adult Education: Possibilities for Transformative Action*. London: Zed Books.

McCann, Laurenellen. 2015. *Experimental Modes of Civic Engagement in Civic Tech: Meeting People Where They Are*. Chicago: Smart Chicago Collaborative.

McDowell, Ceasar, and Melissa Y. Chinchilla. 2016. "Partnering with Communities and Institutions." In *Civic Media: Technology, Design, Practice*, edited by Eric Gordon and Paul Mihailidis, 461–480. Cambridge, MA: MIT Press.

McGregor, Stella, Risa Horn, Peter Kirschmann, Eve Ewing, Sofia Campos, Saul Tannenbaum, Emily Yang, et al. 2013. *The Emancipated City: Reimagining Boston (Urban Myths and the Dream Machine)*. Civic Media Collaborative Design Studio. Cambridge, MA: MIT. https://docs.google.com/document/d/1l368FTU41xmQg-H0bYrHGgxUTLpHR9XGCORWOyXYolg.

Melendez, Steven. 2014. "Contratados: A Yelp to Help Migrant Workers Fight Fraud." *Fast Company*, October 9, 2014. https://www.fastcompany.com/3036812/contratados-is-a-yelp-that-fights-fraud-for-migrant-workers.

Merton, Robert K. 1968. "Continuities in the Theory of Social Structure and Anomie." In *Social Theory and Social Structure*, 215–248. New York: Free Press.

Metz, Cade. 2016. "Forget Apple vs. the FBI: WhatsApp Just Switched on Encryption for a Billion People." *Wired*, April 5, 2016. https://www.wired.com/2016/04/forget-apple-vs-fbi-whatsapp-just-switched-encryption-billion-people.

Mihal, Ivana Julieta. 2014. "Inclusión digital y gestión cultural en el Mercosur: el Programa Puntos de Cultura."

Mikhak, Bakhtiar, Christopher Lyon, Tim Gorton, Neil Gershenfeld, Caroline McEnnis, and Jason Taylor. 2002. "Fab Lab: An Alternate Model of ICT for Development." In *2nd International Conference on Open Collaborative Design for Sustainable Innovation*, pp. 1–7. http://cba.mit.edu/docs/papers/02.00.mikhak.pdf.

Milan, Stefania. 2013. *Social Movements and Their Technologies: Wiring Social Change*. New York: Palgrave Macmillan.

Mill, John Stuart. 1869. *On Liberty*. Longmans, Green, Reader, and Dyer.

Miller, Meg. 2017. "AirBnB Debuts a Toolkit for Inclusive Design." *Fast Company*, August 1, 2017. https://www.fastcodesign.com/90135013/airbnb-debuts-a-toolkit-for-inclusive-design.

Mills, Charles W. 2015. "Race and Global Justice." In *Domination and Global Political Justice*, edited by Barbara Buckinx, Jonathan Trejo-Mathys, and Timothy Waligore, 193–217. New York: Routledge.

Miner, Adam S., Arnold Milstein, Stephen Schueller, Roshini Hegde, Christina Mangurian, and Eleni Linos. 2016. "Smartphone-Based Conversational Agents and Responses to Questions about Mental Health, Interpersonal Violence, and Physical Health." *JAMA Internal Medicine* 176, no. 5: 619–625.

Mohammad, Insiyah, Jason Ma, Husayn Karimi, Lor Holmes, Ben Melançon, and Maya Gaul. 2016. *CERO: Experiments in Marketing and Sales*. Civic Media Collaborative Design Studio. Cambridge, MA: MIT. https://docs.google.com/document/d/1_9unGIc9Pzr3QUBywzrWbviwTRT8I0yLQri2jg08Vq4.

Mohanty, Chandra Talpade. 2013. "Transnational Feminist Crossings: On Neoliberalism and Radical Critique." *Signs: Journal of Women in Culture and Society* 38, no. 4: 967–991.

Molteni, Megan, and Adam Rogers. 2017. "The Actual Science of James Damore's Google Memo." *Wired*, August 15, 2017. https://www.wired.com/story/the-pernicious-science-of-james-damores-google-memo.

Moore, Kelly. 2009. *Disrupting Science: Social Movements, American Scientists, and the Politics of the Military, 1945–1975*. Princeton, NJ: Princeton University Press.

Mouzannar, Hussein, Mesrob I. Ohannessian, and Nathan Srebro. 2019. *From Fair Decision Making to Social Equality*. In *Proceedings of the Conference on Fairness, Accountability, and Transparency (FAT* '19)*, 359–368. New York: ACM. https://doi.org/10.1145/3287560.3287599.

Muller, Michael J. 2003. "Participatory Design: The Third Space in HCI." *Human-Computer Interaction: Development Process* 4235:165–185.

Muñoz, Cecilia, Megan Smith, and D. J. Patil. 2016. *Big Data: A Report on Algorithmic Systems, Opportunity, and Civil Rights*. Executive Office of the President, May 2016.

https://obamawhitehouse.archives.gov/sites/default/files/microsites/ostp/2016_0504_data_discrimination.pdf.

National Center for Women & Information Technology. 2018. *NCWIT's Women in IT: By the Numbers*. NCWIT, April 9, 2018. www.ncwit.org/bythenumbers.

Nelson, Alondra, Thuy Linh Nguyen Tu, and Alicia Headlam Hines, eds. 2001. *Technicolor: Race, Technology, and Everyday Life*. New York: NYU Press.

Neuhart, John, Charles Eames, Ray Eames, and Marilyn Neuhart. 1989. *Eames Design: The Work of the Office of Charles and Ray Eames*. New York: Harry N. Abrams. Translated by José M. Allard S. http://www.eamesoffice.com/the-work/design-q-a-text.

New York Civil Liberties Union. 2014. "Victory in Unlawful Mass Arrest during 2004 RNC the Largest Protest Settlement in History." New York Civil Liberties Union, January 15, 2014. https://www.nyclu.org/en/press-releases/victory-unlawful-mass-arrest-during-2004-rnc-largest-protest-settlement-history.

Nielsen, Lene. 2012. *Personas: User Focused Design*. London: Springer Science & Business Media.

Noble, Safiya Umoja. 2018. *Algorithms of Oppression: How Search Engines Reinforce Racism*. New York: NYU Press.

Norman, Donald A. 2006. *The Design of Everyday Things*. New York: Basic Books.

Norman, Donald A., and Stephen W. Draper. 1986. *User Centered System Design: New Perspectives on Human-Computer Interaction*. Hillsdale, NJ: Lawrence Erlbaum Associates Inc.

Nucera, Diana, Janel Yamishiro, Nina Bianchi, Andy Gunn, and Sasha Costanza-Chock. 2012. "Media-A-Go-Go Hands-on Technology Practice Space." Allied Media Conference 2012 program, 19. Detroit: Allied Media Projects. https://www.alliedmedia.org/sites/tmpstage.dev.altissima.theworkdept.com/files/amc_program2012v6_061212.pdf.

Obama, Barack. 2016. "Giving Every Student an Opportunity to Learn through Computer Science for All." Office of the Press Secretary, January 30, 2016. https://obamawhitehouse.archives.gov/the-press-office/2016/01/30/weekly-address-giving-every-student-opportunity-learn-through-computer.

O'Neil, Daniel X. 2013. "Building a Smarter Chicago." In *Beyond Transparency: Open Data and the Future of Civic Innovation*, edited by Brett Goldstein and Lauren Dyson, 27–38. San Francisco: Code for America Press.

O'Neil, Daniel X. 2016. "An Education in Civic Technology." *Civicist*, June 1, 2016. http://civichall.org/civicist/an-education-in-community-technology.

Open MIC. 2017. *Breaking the Mold: Investing in Racial Diversity in Tech.* Open MIC, February 2017. http://breakingthemold.openmic.org.

Oliver, Mike. 2013. "The Social Model of Disability: Thirty Years On."*Disability & Society* 28, no. 7: 1024–1026. https://doi.org/10.1080/09687599.2013.818773.

Oremus, Will. 2014. "Here Are All the Different Genders You Can Be on Facebook." *Slate*, February 13, 2014. http://www.slate.com/blogs/future_tense/2014/02/13/facebook_custom_gender_options_here_are_all_56_custom_options.html.

Oudshoorn, Nelly, Els Rommes, and Marcelle Stienstra. 2004. "Configuring the User as Everybody: Gender and Design Cultures in Information and Communication Technologies." *Science, Technology, & Human Values* 29, no. 1: 30–63. http://journals.sagepub.com/doi/pdf/10.1177/0162243903259190.

Palfrey, John. 2015. *BiblioTech: Why Libraries Matter More than Ever in the Age of Google.* New York: Basic Books.

Pant, Anupum. 2015. "Racist Sinks." *AweSci-Science Everyday*, September 6, 2015.

Papanek, Victor. 1974. *Design for the Real World.* St. Albans: Paladin.

Papert, Seymour, and Idit Harel. 1991. "Situating Constructionism." *Constructionism* 36, no. 2: 1–11. http://namodemello.com.br/pdf/tendencias/situatingconstrutivism.pdf.

Partridge, Christopher. 2010. *Dub in Babylon: Understanding the Evolution and Significance of Dub Reggae in Jamaica and Britain from King Tubby to Post-Punk.* London: Equinox Publishing.

Patel, Raj. 2009. "Food Sovereignty." *Journal of Peasant Studies* 36, no. 3: 663–706.

Paul, Kari. 2016. "Microsoft Had to Suspend Its AI Chatbot After It Veered into White Supremacy." *Motherboard*, March 24, 2016. https://motherboard.vice.com/en_us/article/kb7zdw/microsoft-suspends-ai-chatbot-after-it-veers-into-white-supremacy-tay-and-you.

Penny, Laurie. 2014. "A Tale of Two Cities: How San Francisco's Tech Boom Is Widening the Gap between Rich and Poor." *New Statesman*, April 9, 2014. http://www.newstatesman.com/laurie-penny/2014/04/tale-two-cities-how-san-franciscos-tech-boom-widening-gap-between-rich-and-poor.

Piepzna-Samarasinha, Leah Lakshmi. 2018. *Care Work: Dreaming Disability Justice.* Vancouver: Arsenal Pulp Press.

Polanyi, Michael. 1958. *Personal Knowledge: Towards a Post-Critical Philosophy.* Chicago: University of Chicago Press.

Prasad, Urvashi. 2012. "Will Reinventing the Toilet Solve the Global Sanitation Crisis?" *Michael & Susan Dell Foundation* (blog), June 21, 2012. https://web.archive

.org/web/20121118070500/https://www.msdf.org/blog/2012/06/wil-reinventing-the-toilet-wont-solve-global-sanitation-crisis.

Prashad, Vijay. 2013. *The Poorer Nations: A Possible History of the Global South.* London: Verso Books.

Purao, Sandeep, John Bagby, and Karthikeyan Umapathy. 2008. "Standardizing Web Services: Overcoming 'Design by Committee.'" In *2008 IEEE Congress on Services: Part I*, 223–230. IEEE. https://doi.org/10.1109/services-1.2008.24.

Pursell, Carroll. 1993. "The Rise and Fall of the Appropriate Technology Movement in the United States, 1965–1985." *Technology and Culture* 34, no. 3 (July): 629–637. https://doi.org/10.2307/3106707.

Racin, Liat, and Eric Gordon. 2018. "Community-Academic Partnerships: The Limitations of Collaborative Research and Overcoming Challenges." Boston Civic Media Consortium. https://elabhome.blob.core.windows.net/resources/mou.pdf.

Raji, Deborah, and Joy Buolamwini. 2019. "Actionable Auditing: Investigating the Impact of Publicly Naming Biased Performance Results of Commercial AI Products." Paper presented at the AAAI/ACM Conference on AI Ethics and Society, Honolulu, HI, January 27–28, 2019.

Raymond, Eric. 1999. "The Cathedral and the Bazaar." *Knowledge, Technology & Policy* 12, no. 3: 23–49.

Reich, Justin, and Mimi Ito. 2017. "From Good Intentions to Real Outcomes: Equity by Design in Learning Technologies." DML Research Hub, October 30, 2017. https://dmlhub.net/publications/good-intentions-real-outcomes-equity-design-learning-technologies.

Reinecke, Katharina, and Abraham Bernstein. 2011. "Improving Performance, Perceived Usability, and Aesthetics with Culturally Adaptive User Interfaces." *ACM Transactions on Computer-Human Interaction (TOCHI)* 18, no. 2, article 8: 1–29. New York: ACM. https://dl.acm.org/citation.cfm?doid=1970378.1970382.

Renzi, Alessandra. 2015. "Info-Capitalism and Resistance: How Information Shapes Social Movements." *Interface: A Journal on Social Movements* 7, no. 2: 98–119. http://www.interfacejournal.net/wordpress/wp-content/uploads/2015/12/Issue-7-2-Renzi.pdf.

Resnick, Brian. 2017. "The March for Science Is Forcing Science to Reckon with Its Diversity Problem." *Vox*, March 24, 2017. https://www.vox.com/science-and-health/2017/3/24/15028396/march-for-science-diversity.

Resnick, Mitchel. 2017. *Lifelong Kindergarten: Cultivating Creativity through Projects, Passion, Peers, and Play*. Cambridge, MA: MIT Press.

Resnick, Mitchel, John Maloney, Andrés Monroy-Hernández, Natalie Rusk, Evelyn Eastmond, Karen Brennan, Amon Millner, et al. 2009. "Scratch: Programming for All." *Communications of the ACM* 52, no. 11 (November): 60–67. https://doi.org/10.1145/1592761.1592779.

Resnick, Mitchell, and Natalie Rusk. 1996. "Computer Clubhouses in the Inner City: Access Is Not Enough." *American Prospect* 7, no. 27: 60–68.

Resnick, Mitchell, Natalie Rusk, and Stina Cooke. 1998. "The Computer Clubhouse: Technological Fluency in the Inner City." In *High Technology and Low-Income Communities: Prospects for the Positive Use of Advanced Information Technology*, edited by Donald Schön, Bishwapriya Sanyal, and William J. Mitchell, 263–285. Cambridge, MA: MIT Press.

Rhode, Deborah L. 1991. *Justice and Law*. Brighton, MA: Harvard University Press.

Richard, Gabriela, Yasmin B. Kafai, Barrie Adleberg, and Orkan Telhan. 2015. "StitchFest: Diversifying a College Hackathon to Broaden Participation and Perceptions in Computing." In *SIGCSE 2015—Proceedings of the 46th ACM Technical Symposium on Computer Science Education*, edited by Carl Alphonce, Adrienne Decker, Kurt Eiselt, and Jodi Tims, 114–119. New York: ACM. https://dl.acm.org/citation.cfm?doid=2676723.2677310.

Ries, Eric. 2011. *The Lean Startup: How Today's Entrepreneurs Use Continuous Innovation to Create Radically Successful Businesses*. New York: Crown Business.

Riggs, William M., and Eric Von Hippel. 1994. "The Impact of Scientific and Commercial Values on the Sources of Scientific Instrument Innovation." *Research Policy* 23, no. 4: 459–469.

Robinson, Pamela J., and Peter A. Johnson. 2016. "Civic Hackathons: New Terrain for Local Government-Citizen Interaction?" *Urban Planning* 1, no. 2: 65–74.

Rodriguez, Clemencia. 2001. *Fissures in the Mediascape: An International Study of Citizens' Media*. Cresskill, NJ: Hampton Press.

Rogers, Everett M. 1962. *Diffusion of Innovations*. Glencoe, IL: Free Press of Glencoe.

Rose, Tricia. 1994. *Black Noise: Rap Music and Black Culture in Contemporary America*. Vol. 6. Middletown, CT: Wesleyan University Press.

Ross, Andrew, ed. 1997. *No Sweat: Fashion, Free Trade, and the Rights of Garment Workers*. London: Verso.

Ross, Joel, Lilly Irani, M. Six Silberman, Andrew Zaldivar, and Bill Tomlinson. 2010. "Who Are the Crowdworkers? Shifting Demographics in Mechanical Turk." In *CHI'10 Extended Abstracts on Human Factors in Computing Systems*, 2863–2872. New York: ACM. http://citeseerx.ist.psu.edu/viewdoc/download;jsessionid=B496D21FD15C84E81F1746CEA516017C?doi=10.1.1.187.8270.

Roston, Michael. 2017. "The March for Science: Why Some Are Going, and Some Will Sit Out." *New York Times*, April 17, 2017. https://www.nytimes.com/2017/04/17/science/march-for-science-voices.html.

Routledge, Paul. 2003. "Convergence Space: Process Geographies of Grassroots Globalization Networks." *Transactions of the Institute of British Geographers* 28, no. 3: 333–349.

Rutkin, A. 2016. "Digital discrimination." *New Scientist* 231, no. 3084: 18–19. https://doi.org/10.1016/S0262-4079(16)31364-1.

Sabelli, Nora. 2008. "Constructionism: A New Opportunity for Elementary Science Education." *DRL Division of Research on Learning in Formal and Informal Settings*, 193–206.

Sadat, Leila Nadya. 2005. "Ghost Prisoners and Black Sites: Extraordinary Rendition under International Law." *Case Western Reserve Journal of International Law* 37:309–342.

Sadedin, Suzanne. 2017. "A Scientist's Take on the Biological Claims from the Infamous Google Anti-diversity Manifesto." *Forbes*. https://www.forbes.com/sites/quora/2017/08/10/a-scientists-take-on-the-biological-claims-from-the-infamous-google-anti-diversity-manifesto/#4b75d9dd3364.

Sanders, Elizabeth B-N., and Pieter Jan Stappers. 2008. "Co-creation and the New Landscapes of Design." *CoDesign* 4, no. 1: 5–18.

Sanoff, Henry. 2008. "Multiple Views of Participatory Design." *International Journal of Architectural Research* 2, no. 1: 57–69.

Santa Ana, Otto, Layza López, and Edgar Munguía. 2010. "Framing Peace as Violence: U.S. Television News Depictions of the 2007 Los Angeles Police Attack on Immigrant Rights Marchers." *Aztlán* 35, no. 1: 69–101.

Schmider, Alex. 2016. "GLAAD Works with Tinder to Create Trans-Inclusive App Update." GLAAD, November 14, 2016. https://www.glaad.org/blog/glaad-works-tinder-create-trans-inclusive-app-update.

Schneier, Bruce. 2006. *Beyond Fear: Thinking Sensibly about Security in an Uncertain World*. New York: Springer Science & Business Media.

Scholz, Trebor, ed. 2013. *Digital Labor: The Internet as Playground and Factory*. New York: Routledge.

Scholz, Trebor, and Nathan Schneider, eds. 2016. *Ours to Hack and to Own: The Rise of Platform Cooperativism, a New Vision for the Future of Work and a Fairer Internet*. New York: OR Books.

Schön, D. A. 1983. *The Reflective Practitioner: How Professionals Think in Action*. New York: Basic Books.

Schön, Donald A. 1987. *Educating the Reflective Practitioner: Toward a New Design for Teaching and Learning in the Professions*. San Francisco: Jossey-Bass.

Schudson, Michael. 1998. *The Good Citizen*. New York: Free Press.

Schuler, Douglas, and Aki Namioka, eds. 1993. *Participatory Design: Principles and Practices*. Boca Raton, FL: CRC Press.

Schulz, Nick. 2006. "The Crappiest Invention of All Time: Why the Auto-Flushing Toilet Must Die." *Slate*. https://slate.com/culture/2006/03/the-crappiest-invention-of-all-time.html.

Schumacher, Ernst Friedrich. 1999. *Small Is Beautiful: Economics as If People Mattered; 25 Years Later … with Commentaries*. Point Roberts, WA: Hartley & Marks.

Science for the People. 2018a. "Pickets at Microsoft to Support Workers' Demand: End the ICE Contract!" Science for the People, June 25, 2018. https://sciencforthepeople.org/wp-content/uploads/2018/06/SftP-Microsoft-Press-Release.pdf.

Science for the People. 2018b. "Science for the People Documentary." YouTube, July 30, 2018. https://www.youtube.com/watch?=263&v=BOxn_cXfNsM.

Segarra, Lisa Marie. 2018. "More Than 20,000 Google Employees Participated in Walkout Over Sexual Harassment Policy." *Fortune*, November 3, 2018. http://fortune.com/2018/11/03/google-employees-walkout-demands.

Segarra, Lisa Marie, and David Johnson. 2017. "Find Out If President Trump Would Let You Immigrate to America." *TIME*, August 7, 2017. http://time.com/4887574/trump-raise-act-immigration.

Selvaraj, Shivaani Aruna. 2016. *The Quest for Really Useful Knowledge: An Institutional Ethnography of Community Adult Education in the Digital Age*. DEd diss, Pennsylvania State University.

Shadduck-Hernández, Janna, Zacil Pech, Mar Martinez, and Marissa Nuncio. *Dirty Threads, Dangerous Factories: Health and Safety in Los Angeles' Fashion Industry*. Garment Worker Center, December 2016. http://garmentworkercenter.org/wp-content/uploads/2016/12/DirtyThreads.pdf.

Shah, Silky, Carly Perez, Sean Flynn, Pia Zaragoza, Chrislene DeJean, and Marcelo Nieves. 2014. *Bedtime Stories: An Interactive Documentary about the Immigrant Detention System*. Civic Media Collaborative Design Studio. Cambridge, MA: MIT. https://docs.google.com/document/d/1vpf0ehxmv0qxXu0dJ-VX-eoBwACsU7A1Ih4iReSDpC4.

Shepard, Benjamin, and Ronald Hayduk, eds. 2002. *From ACT UP to the WTO: Urban Protest and Community Building in the Era of Globalization*. London: Verso.

Shetterly, Margot Lee. 2017. *Hidden Figures: The Untold Story of African-American Women Who Helped to Win the Space Race*. New York: Harper Collins, 2017.

Sifry, Micah L. 2012. "From TXTMob to Twitter: How an Activist Tool Took Over the Conventions." TechPresident, August 25, 2012. http://techpresident.com/news/22775/txtmob-twitter-how-activist-tool-took-over-conventions.

Silbey, Jessica. 2014. *The Eureka Myth: Creators, Innovators, and Everyday Intellectual Property*. Stanford, CA: Stanford University Press.

Silbey, Susan S. 2018. "#MeToo at MIT: Harassment and Systemic Gender Subordination." *MIT Faculty Newsletter*, January/February 2018. http://web.mit.edu/fnl/volume/303/silbey.html.

Siles, Ignacio. 2013. "Inventing Twitter: An Iterative Approach to New Media Development." *International Journal of Communication* 7:23.

Simon, Herbert A. 1996. *The Sciences of the Artificial*. Cambridge, MA: MIT Press.

Skinner, David. 2006. "The Age of Female Computers." *New Atlantis* 12 (Spring): 96–103.

Smith, Adrian, Adrian Ely, Mariano Fressoli, Dinesh Abrol, and Elisa Arond. 2016. *Grassroots Innovation Movements*. New York: Routledge.

Smith, Linda Tuhiwai. 2013. *Decolonizing Methodologies: Research and Indigenous Peoples*. London: Zed Books Ltd.

Smith, Matt, and Aura Bogado. 2018. "Immigrant Children Forcibly Injected with Drugs, Lawsuit Claims." *Reveal*, June 20, 2018. https://www.revealnews.org/blog/immigrant-children-forcibly-injected-with-drugs-lawsuit-claims.

Smyth, Thomas, and Jill Dimond. 2014. "Anti-oppressive Design." *Interactions* 21, no. 6: 68–71.

Soderberg, Johann. 2013. "Determining Social Change: The Role of Technological Determinism in the Collective Action Framing of Hackers." *New Media & Society* 15, no. 8: 1277–1293.

Solorzano, Daniel, Miguel Ceja, and Tara Yosso. 2000. "Critical Race Theory, Racial Microaggressions, and Campus Racial Climate: The Experiences of African American College Students." *Journal of Negro Education* 69, nos. 1–2 (Winter/Spring): 60–73.

Spelic, Sherri. 2018. "What If? And What's Wrong? Design Thinking and Thinking about Design We Can't Easily See." Medium, January 14, 2018. https://medium.com/@edifiedlistener/what-if-and-whats-wrong-f617ada90216.

Spitzer, Danny. 2016. "The Very First Oakland Co-op DiscoTech." Medium, May 6, 2016. https://medium.com/@daspitzberg/the-very-first-oakland-co-op-discotech-a1d7b471562.

Sreberny, Annabelle. 2004. "WSIS: Articulating Information at the Summit." *International Communication Gazette* 66, nos. 3–4: 193–201.

Srinivasan, Ramesh. 2017. *Whose Global Village? Rethinking How Technology Shapes Our World*. New York: NYU Press.

Starr, Paul. 2004. *The Creation of the Media: Political Origins of Modern Communications*. New York: Basic Books.

Steen, Marc. 2011. "Tensions in Human-Centred Design." *CoDesign* 7, no. 1: 45–60.

Steen, Marc. 2013. "Co-design as a Process of Joint Inquiry and Imagination." *Design Issues* 29, no. 2 (Spring): 16–28.

Stevens, Sean, and Jonathan Haidt. 2017. "The Google Memo: What Does the Research Say About Gender Differences?" *Heterodox Academy*, August 10, 2017. https://heterodoxacademy.org/the-google-memo-what-does-the-research-say-about-gender-differences.

Story, Molly Follette. 2001. "Principles of Universal Design." In *Universal Design Handbook*, edited by Wolfgang F. E. Preiser and Elaine Ostroff, 31–36. New York: McGraw-Hill Professional.

Subcomandante Marcos. 2000. "Zapatista Army of National Liberation Statement on the Kosovo War." *Left Curve* 24:22.

Sue, Derald Wing. 2010. *Microaggressions and Marginality: Manifestation, Dynamics, and Impact*. West Sussex, UK: John Wiley & Sons.

Sue, Derald Wing, Christina M. Capodilupo, Gina C. Torino, Jennifer M. Bucceri, Aisha Holder, Kevin L. Nadal, and Marta Esquilin. 2007. "Racial Microaggressions in Everyday Life: Implications for Clinical Practice." *American Psychologist* 62, no. 4: 271.

Sue, Derald Wing, Annie I. Lin, and David P. Rivera. 2009. "Racial Microaggressions in the Workplace: Manifestation and Impact." In *Praeger Perspectives: Race and Ethnicity in Psychology*, edited by J. L. Chin, 157–172. Santa Barbara, CA: Praeger.

Swift, Mike. 2010. "Blacks, Latinos, and Women Lose Ground at Silicon Valley Tech Companies." *San Jose Mercury News*, February 11, 2010. http://www.mercurynews.com/ci_14383730?source=pkg.

Sydell, Laura. 2018. "Immigrant Rights Group Turns Down $250,000 from Tech Firm over Ties to Border Patrol." NPR, July 19, 2018. https://www.npr.org/2018/07/19/630358800/immigrant-rights-group-turns-down-250-000-from-tech-firm-over-ties-to-border-pat.

Tarrow, Sidney. 2010. "Preface." In *The World Says No to War: Demonstrations against the War on Iraq*, vol. 33, edited by Stefaan Walgrave and Dieter Rucht, vii. Minneapolis: University of Minnesota Press.

Taylor, T. L. 2018. *Watch Me Play: Twitch and the Rise of Game Live Streaming*. Princeton, NJ: Princeton University Press.

ten Oever, Niels. 2018. "Guidelines for Human Rights Protocol Considerations." IETF Human Rights Protocol Considerations Research Group, September 7, 2018. https://datatracker.ietf.org/doc/draft-irtf-hrpc-guidelines.

Terranova, Tiziana. 2000. "Free Labor: Producing Culture for the Digital Economy." *Social Text* 18, no. 2: 33–58.

Thatcher, Margaret. 1987. "Aids, Education and the Year 2000!" Interview by Douglas Keay. *Woman's Own*, October 31, 1987.

Thurm, Scott. 2018. "These Tech Companies Will Need More Women on their Boards." *Wired*, October 2, 2018. https://www.wired.com/story/these-tech-companies-will-need-more-women-on-boards.

Tong, Traci. 2017. "A Lack of Clean and Safe Toilets Leaves Women Vulnerable to Rape and Attacks." PRI, November 29, 2017. https://www.pri.org/stories/2017-11-29/lack-clean-and-safe-toilets-leaves-women-vulnerable-rape-and-attacks.

Toupin, Sophie. 2014. "Feminist Hackerspaces: The Synthesis of Feminist and Hacker Cultures." *Journal of Peer Production*, no. 5. http://peerproduction.net/editsuite/issues/issue-5-shared-machine-shops/peer-reviewed-articles/feminist-hackerspaces-the-synthesis-of-feminist-and-hacker-cultures.

Trans*H4CK. n.d. "Our Mission." Trans*H4CK. Accessed January 14, 2017. http://www.transhack.org/mission.

Treré, Emiliano. 2012. "Social Movements as Information Ecologies: Exploring the Coevolution of Multiple Internet Technologies for Activism." *International Journal of Communication* 6:19.

Trujillo, Josmar. 2004. "CNN Attempts Actual Journalism—but Reverts to Embedded Reporting." *FAIR*, August 19, 2004. https://fair.org/home/cnn-attempts-actual-journalism-but-reverts-to-embedded-reporting.

Truth, Sojourner. 1995. "Ain't I a Woman?" In *With Pen and Voice: A Critical Anthology of Nineteenth-Century African-American Women*. Carbondale: Southern Illinois University Press.

Tufekci, Zeynep. 2017. *Twitter and Tear Gas: The Power and Fragility of Networked Protest*. New Haven, CT: Yale University Press.

Turner, Fred. n.d. "Don't Be Evil." *Logic*, no. 3: 17–37.

Turner, Fred. 2009. "Burning Man at Google: A Cultural Infrastructure for New Media Production." *New Media & Society* 11, nos. 1–2: 73–94.

Turner, Fred. 2010. *From Counterculture to Cyberculture: Stewart Brand, the Whole Earth Network, and the Rise of Digital Utopianism*. Chicago: University of Chicago Press.

Turner, Phil, and Susan Turner. 2011. "Is Stereotyping Inevitable when Designing with Personas?" *Design Studies* 32, no. 1: 30–44.

Tweney, Dylan. 2009. "DIY Freaks Flock to 'Hacker Spaces' Worldwide." *Wired*, March 29, 2009. https://www.wired.com/2009/03/hackerspaces.

Tynes, Brendesha M., Chad A. Rose, and Suzanne L. Markoe. 2013. "Extending Campus Life to the Internet: Social Media, Discrimination, and Perceptions of Racial Climate." *Journal of Diversity in Higher Education* 6, no. 2: 102.

Tyson, Alec, and Shiva Maniam. 2016. "Behind Trump's Victory: Divisions by Race, Gender, Education." Fact Tank, Pew Research Center, November 9, 2016. http://www.pewresearch.org/fact-tank/2016/11/09/behind-trumps-victory-divisions-by-race-gender-education.

Union of the Physically Impaired Against Segregation (UPAIS). 1975. "Fundamental Principles of Disability." *The Disability Studies Archive*. https://disability-studies.leeds.ac.uk/library/author/upias.

Varon, Joana, and Corinne Cath. 2015. "Human Rights Protocol Considerations Methodology." IETF Human Rights Protocol Considerations Research Group, July 6, 2015. https://tools.ietf.org/html/draft-varon-hrpc-methodology-00.

Volz, Dustin. 2017. "IBM Urged to Avoid Working on 'Extreme Vetting' of U.S. Immigrants." Reuters, November 16, 2017. https://www.reuters.com/article/us-ibm-immigration/ibm-urged-to-avoid-working-on-extreme-vetting-of-u-s-immigrants-idUSKBN1DG1VT.

Von Hippel, Eric. 2005. *Democratizing Innovation*. Cambridge, MA: MIT Press.

VozMob Project. 2011. "Mobile Voices: Projecting the Voices of Immigrant Workers by Appropriating Mobile Phones for Popular Communication." In *Communications Research in Action: Scholar-Activist Collaborations for a Democratic Public Sphere*, edited by Philip M. Napoli and Minna Aslama, 177–196. New York: Fordham University Press.

Wachter-Boettcher, Sara. 2017. *Technically Wrong: Sexist Apps, Biased Algorithms, and Other Threats of Toxic Tech*. New York: W. W. Norton & Company.

Wagoner, Maya. 2017. "Technology against Technocracy: Toward Design Strategies for Critical Community Technology." MSc diss., Massachusetts Institute of Technology. https://cmsw.mit.edu/wp/wp-content/uploads/2017/10/Maya-Wagoner-Technology-Against-Technocracy-Toward-Design-Strategies-for-Critical-Community-Technology.pdf.

Wajcman, Judy. 1991. *Feminism Confronts Technology*. University Park, PA: Penn State Press.

Wajcman, Judy. 2010. "Feminist Theories of Technology." *Cambridge Journal of Economics* 34, no. 1: 143–152.

Wakabayashi, Daisuke. 2017. "Google Fires Engineer Who Wrote Memo Questioning Women in Tech." *New York Times*, August 7, 2017. https://www.nytimes.com/2017/08/07/business/google-women-engineer-fired-memo.html.

Walgrave, Stefaan and Dieter Rucht. 2010. "Introduction." In *The World Says No to War: Demonstrations against the War on Iraq*, vol. 33, edited by Stefaan Walgrave and Dieter Rucht, xiii. Minneapolis: University of Minnesota Press.

Wallerstein, Immanuel. 2011. *The Modern World-System I: Capitalist Agriculture and the Origins of the European World-Economy in the Sixteenth Century*. Vol. 1. Berkeley and Los Angeles: University of California Press.

Walter-Herrmann, Julia, and Corinne Büching, eds. 2014. *FabLab: Of Machines, Makers and Inventors*. Bielefeld, Germany: Transcript Verlag.

Watkins, Craig S. 2019. *Don't Knock the Hustle: Young Creatives, Tech Ingenuity, and the Making of a New Innovation Economy*. Boston: Beacon Press.

Waxman, Olivia. 2017. "Women in Tech and the History behind That Controversial Google Diversity Memo." *TIME*, August 8, 2017. http://time.com/4892094/google-diversity-history-memo.

Web We Want. 2014. "III DiscoTech in Istanbul: Sharing Circumvention Tactics." Web We Want, August 22, 2014. https://webwewant.org/news/iii_discotech_in_istanbul_sharing_circumvention_tactics.

Weeden, Kim A., Youngjoo Cha, and Mauricio Bucca. 2016. "Long Work Hours, Part-Time Work, and Trends in the Gender Gap in Pay, the Motherhood Wage Penalty, and the Fatherhood Wage Premium." *RSF: The Russell Sage Foundation Journal of the Social Sciences* 2, no. 4 (August): 71–102.

Weishaar, Kate, Calvin Zhong, and Daniel Cheng. 2017. *Open Book/Libro Abierto*. Civic Media Collaborative Design Studio. Cambridge, MA: MIT. https://docs.google.com/document/d/1fEj3a1byCX-y1dd0mFZVrKCTCFao_vwEnzFV0oc04tU.

White, Sarah C. 1996. "Depoliticising Development: The Uses and Abuses of Participation." *Development in Practice* 6, no. 1: 6–15.

Wiener, Anna. 2017. "How Silicon Valley's Workplace Culture Produced James Damore's Google Memo." *New Yorker*, August 10, 2017. https://www.newyorker.com/tech/elements/how-silicon-valleys-workplace-culture-produced-james-damores-google-memo.

Williamson, Bess. 2019. *Accessible America: A History of Disability and Design*. New York: NYU Press.

Williamson, Sara Elizabeth. 2011. "The Right to Design Disability and Access in the United States, 1945–1990." PhD diss., University of Delaware.

Willis, Anne-Marie. 2006. "Ontological Designing." *Design Philosophy Papers* 4, no. 2: 69–92.

Willoughby, Kelvin W. 1990. *Technology Choice: A Critique of the Appropriate Technology Movement.* Boulder, CO: Westview Press.

Wilson, Brent Gayle. 1996. *Constructivist Learning Environments: Case Studies in Instructional Design.* Englewood Cliffs, NJ: Educational Technology Publications.

Wilson, Franklin D. 2016. *Generational Changes in Racial Inequality in Occupational Attainment, 1950–2010: A Synthetic Cohort Analysis.* University of Wisconsin-Madison, Institute for Research on Poverty.

Wilson, Valerie, and William M. Rodgers III. 2016. "Black-White Wage Gaps Expand with Rising Wage Inequality." Economic Policy Institute, September 20, 2016. https://www.epi.org/publication/black-white-wage-gaps-expand-with-rising-wage-inequality.

Winner, Langdon. 1980. "Do Artifacts Have Politics?" *Daedalus* 109, no. 1 (Winter): 121–136.

Wittkower, D. E. 2016. "Principles of Anti-discriminatory Design." *Philosophy Faculty Publications*, no. 28. https://digitalcommons.odu.edu/philosophy_fac_pubs/28.

Wolfson, Todd. 2014. *Digital Rebellion: The Birth of the Cyber Left.* Champaign: University of Illinois Press.

Wood, Lesley J. 2014. *Crisis and Control: The Militarization of Protest Policing.* London: Pluto Press.

Woods, Jamila. 2016. "Way Up." Recorded 2016. Closed Sessions. Track 13 on *HEAVN.* https://soundcloud.com/jamilawoods/way-up-prod-by-oddcouple.

Wu, Kathy, Sam Daitzman, Aditi Jhaveri, Jade Bacherman, Maxwell Kreppein, Nina Cragg, and Alex Jin. 2017. *Rainbow: Central Square Collage.* Civic Media Collaborative Design Studio. Cambridge, MA: MIT. https://docs.google.com/document/d/1fpR5GLREKT6m4jd4QQMMq-q2UdnuAIKzXIT68o29axU.

Zaleski, Katharine. 2017. "The Maddeningly Simple Way Tech Companies Can Employ More Women." *New York Times*, August 15, 2017. https://www.nytimes.com/2017/08/15/opinion/silicon-valley-women-hiring-diversity.html.

Zuboff, Shoshana. 2015. "Big Other: Surveillance Capitalism and the Prospects of an Information Civilization." *Journal of Information Technology* 30, no. 1: 75–89.

Zukin, Sharon, and Max Papadantonakis. 2017. "Hackathons as Co-optation Ritual: Socializing Workers and Institutionalizing Innovation in the 'New' Economy." In *Precarious Work*, vol. 31, Research in the Sociology of Work, edited by Arne L. Kalleberg and Steven P. Vallas, 157–181. Bingley, UK: Emerald Publishing Limited.

Index

The letter *f* following a page number denotes a figure.